PSPICE
and
MATLAB®
for Electronics

An Integrated Approach

VLSI CIRCUITS SERIES

Series Editor

Wai-Kai Chen

PUBLISHED TITLES

PSPICE and MATLAB® for Electronics: An Integrated Approach,
John O. Attia

VLSI Design,
M. Michael Vai

FORTHCOMING TITLES

Analog VLSI Design Automation,
Sina Balkir, Gunhan Dundar, and Selcuk Ogrenci

PSPICE
and
MATLAB®
for Electronics
An Integrated Approach

JOHN O. ATTIA

CRC PRESS

Boca Raton   London   New York   Washington, D.C.

Library of Congress Cataloging-in-Publication Data

Attia, John Okyere.
    PSPICE and MATLAB for electronics: an integrated approach / John Okyere Attia.
        p. cm. -- (VLSI circuits series)
    Includes bibliographical references and index.
    ISBN 0-8493-1263-9 (alk. paper)
    1. Integrated circuits, Very large scale integration--Design and construction--Data
processing. 2. PSPICE. 3. MATLAB. 4. Electronic circuit design--Data Processing. I.
Title. II. Series

TK7874.75 .A88 2002
621.39'5'--dc21   2002017433

Visit the CRC Press Web site at www.crcpress.com

No claim to original U.S. Government works
International Standard Book Number 0-8493-1263-9
Library of Congress Card Number 2002017433
Printed in the United States of America 1 2 3 4 5 6 7 8 9 0
Printed on acid-free paper

To my parents

for

their unfailing love and encouragement

Preface

SPICE is an industry-standard software for circuit simulation. It can perform dc, ac, transient, Fourier, and Monte Carlo analyses. In addition, SPICE has device models incorporated into its package. Furthermore, there is an extensive library of device models available that a SPICE user can employ for simulation and design. PSPICE, a SPICE package by Cadence Design, has an analog behavioral model facility that allows modeling of analog circuit functions using mathematical equations, tables, and transfer functions. The above features of PSPICE are unmatched by other scientific packages.

MATLAB® is primarily a tool for matrix computations. It has numerous functions for data processing and analysis. In addition, there is a rich set of plotting capabilities integrated into the MATLAB package. Furthermore, because MATLAB is also a programming environment, a user can extend the MATLAB functional capabilities by writing new modules (m-files).

This book uses the strong features of PSPICE and the powerful functions of MATLAB for electronic circuit analysis. PSPICE can be used to perform dc, ac, transient, Fourier, temperature, and Monte Carlo analysis of electronic circuits with device models and subsystem sub-circuits. Then, MATLAB can be used to perform calculation of device parameters, curve fitting, numerical integration, numerical differentiation, statistical analysis, and two- and three-dimensional plots.

PSPICE has the postprocessor package PROBE, which can be used for plotting PSPICE results. In addition, PROBE also has built-in functions that can be used to do simple signal processing. However, the PROBE functions are extremely limited compared to those of MATLAB.

The goals of writing this book are (1) to provide the reader with an introduction to PSPICE; (2) to provide the reader with a simple, easy, hands-on introduction to MATLAB; and (3) to demonstrate the combined power of PSPICE and MATLAB for solving electronics problems.

This book is unique. It is the first time a book is being written that provides an introduction to both MATLAB and PSPICE. In addition, it is the first time a book is being written that integrates the strong features of PSPICE and the powerful functions of MATLAB for problem solving in electronics.

Audience

The book can be used by students, professional engineers, and technicians. Section I (Chapters 1 and 2) is a basic introduction to the PSPICE software program. Section II (Chapters 3 and 4) of the book can be used as a primer for MATLAB. It will be useful to all students and professionals who want a basic introduction to MATLAB. Section III (Chapters 5, 6, and 7) is for electrical and electrical engineering technology students and professionals who want to use both PSPICE and MATLAB to explore the characteristics of semiconductor devices and to apply the two software packages for analysis of electronic circuits and systems.

Organization

The book is divided into three sections: Section I provides an introduction to PSPICE. The basic PSPICE commands are discussed in Chapter 1 and the advanced features of PSPICE are covered in Chapter 2. The chapters have several examples to illustrate the application of PSPICE in electronics circuit analysis.

Section II is an introduction to MATLAB. Circuit analysis and electronics applications using MATLAB are explored. It is recommended that the reader work through and experiment with the examples at a computer while reading Chapters 1 through 4. A hands-on approach is one of the best ways of learning PSPICE and MATLAB.

Section III includes Chapters 5, 6, and 7. The topics discussed in this section are diodes, operational amplifiers, and transistor circuits. Applications of PSPICE and MATLAB for problem solving in electronics are discussed. Extensive examples reveal the combined power of PSPICE and MATLAB for solving problems in electronics. Each chapter has its own bibliography and problems.

Examples at CRC Press Web Site

The text contains a large number of PSPICE and MATLAB examples. The soft copies of the PSPICE programs, data files, and MATLAB m-files of the examples in the book are available at the CRC Press Web site (www.crcpress.com). The reader can run the examples without having to enter the commands. The examples can also be modified to suit the needs of the reader.

Acknowledgments

I appreciate the suggestions and comments made by Penrose Cofie in reviewing this book. I am grateful to Monica Bibbs for typing the manuscript, and to Julian Farquharson and Rodrigo Lozano for drawing the circuit diagrams found in the text. Special thanks go to Nora Konopka, acquisitions editor at CRC Press. Special thanks also go to Helena Redshaw, supervisor, EPD department at CRC Press.

MATLAB® is a registered trademark of the MathWorks, Inc. For MATLAB product information, ase contact: The MathWorks, Inc., 3 Apple Hill Drive, Natick, MA 01760-2098 USA; Phone: 3-647-7000; Fax: 508-647-7001; E-mail: info@mathworks.com; Web: www.mathworks.com

Author

John Okyere Attia, Ph.D., is professor and head of the Electrical Engineering Department at Prairie View A&M University, as well as the associate director for the NASA Center for Applied Radiation Research. He teaches graduate and undergraduate courses in electrical engineering in the field of electronics, circuit analysis, instrumentation systems, digital signal processing and VLSI Design. Dr. Attia has been teaching for the past 20 years.

Dr. Attia earned a Ph.D. in electrical engineering from the University of Houston, an M.S. from the University of Toronto, and a B.S. from the University of Science and Technology. He worked briefly at AT&T Bell Laboratories and 3M. Dr. Attia has more than 50 publications. His research interests include innovative electronic circuit designs for radiation environment, and radiation testing of electronic devices, circuits and systems. Dr. Attia is the author of *Electronics and Circuits Analysis Using MATLAB* (CRC Press, 1999).

Dr. Attia has received several honors. He has twice received the Outstanding Teacher's Award. He is a member of the following honor societies: Sigma Xi, Tau Beta Pi, Kappa Alpha Kappa, and Eta Kappa Nu. Dr. Attia is a registered Professional Engineer in the state of Texas, and a Senior Member of the Institute of Electrical and Electronics Engineers (IEEE).

Contents

Section II   MATLAB Primer

List of Solved Examples

Chapter 1

Chapter 2

Chapter 3

Chapter 4

Chapter 5

Chapter 6

Chapter 7

Section I

Introduction to PSPICE

1

PSPICE Fundamentals

1.1 Introduction

SPICE (Simulated Program with Integrated Circuit Emphasis) is an industry-standard software for circuit simulation. It can be used among other circuit analysis to perform alternating current, direct current, Fourier, and Monte Carlo analyses. SPICE has continued to be the standard for analog circuit simulation for the electronics industry over the past two decades. There are several SPICE-derived simulation packages; among these are Orcad PSPICE, Meta-software HSPICE, and Intusoft IS-SPICE.

PSPICE has additional features as compared to classical SPICE. Among some of the useful features are

1. PSPICE has a post-processor program, PROBE, which can be used for interactive graphical display of simulation results.
2. Current flowing through an inductor, capacitor, or resistor can be easily obtained without inserting a current monitor in series with the passive elements.
3. PSPICE has analog behavioral model facility that allows modeling of analog circuit functions using mathematical equations, tables, and transfer functions.
4. PSPICE does not distinguish between uppercase and lowercase characters. In SPICE, all characters in the source file must be uppercase. (For example, rab and RAB are considered equivalent in PSPICE).

A general SPICE program consists of the following components:

- Title
- Element statements
- Control statements
- End statements

The following two sections discuss the element and control statements.

1.2   Element Statements

The element statement specifies the elements in the circuit. The element statement contains the (1) element name, (2) the circuit nodes to which each element is connected, and (3) the values of the parameters that electrically characterize the element.

The element name must begin with a letter of the alphabet that is unique to a circuit element, source, or subcircuit. Table 1.1 shows the beginning alphabet of an element name and the corresponding element.

Circuit nodes are positive integers. The nodes in the circuit need not be numbered sequentially. Node 0 is predefined for the ground of a circuit. To prevent error messages, all nodes must be connected to at least two elements.

Element values can be integer, floating pointer number, integer floating point followed by an exponent, or floating point or integer followed by scaling factors shown in Table 1.2.

Any character after the scaling factor abbreviation is ignored in SPICE. For example, a 5000-Ohm resistor can be written as 5000, 5000.00Ohm, 5K, 5E3, 5Kohm, or 5KR.

The element statements of some common elements such as a resistor, inductor, capacitor, independent voltage source, and independent current source will now be described.

TABLE 1.1

Element Name and Corresponding Element

First Letter of Element Name	Circuit Element, Sources, and Subcircuit
B	GaAs field-effect transistor
C	Capacitor
D	Diode
E	Voltage-controlled voltage source
F	Current-controlled current source
G	Voltage-controlled current source
H	Current-controlled voltage source
I	Independent current source
J	Junction field-effect transistor
K	Mutual inductors (transformers)
L	Inductor
M	MOS field-effect transistor
Q	Bipolar junction transistor
R	Resistor
S	Voltage-controlled switch
T	Transmission line
V	Independent voltage source
X	Subcircuit

TABLE 1.2

Abbreviations of SPICE Scaling Factor

Suffix Letter	Metric Prefix	Multiplying Factor
T	Tera	10^{12}
G	Giga	10^9
Meg	Mega	10^6
K	Kilo	10^3
M	Milli	10^{-3}
U	Micro	10^{-6}
N	Nano	10^{-9}
P	Pico	10^{-12}
F	Femto	10^{-15}
Mil	Millimeter	25.4×10^{-6}

Resistors

The general format for describing resistors is

Rname N+ N− value [TC = TC1,TC2]

where

The name must start with the letter **R**.

N+ and **N−** are the positive and negative nodes of the resistor. Conventional current flows from the positive node N+ through the resistor to the negative node N−.

value specifies the value of the resistor. The latter may be positive or negative, but not zero.

TC1 and **TC2** are the temperature coefficients. The default values are zero. If they are nonzero, then the resistance is given by the formula

$$\text{Resistor value} = value\left[1 + TC1\left(T - T_{nom}\right) + TC2\left(T - T_{nom}\right)^2\right] \quad (1.1)$$

where

$TC1$ is linear temperature coefficient,

$TC2$ is quadrature temperature coefficient,

T_{nom} is the nominal temperature, set using TNOM option; its default value is 27°C.

Inductors

The general format for describing linear inductors is

Lname N+ N− value [IC = initial_current]

where
> The inductor name must start with the letter **L**.
> **N+** and **N−** are positive and negative nodes of the inductor, respectively. Conventional current flows from the positive node to the negative node.
> **value** specifies the values of the inductance.
> The initial condition for transient analysis is assigned using **IC =** initial_current to specify the initial current.

Capacitors

The general format for describing linear capacitors is

> **Cname N+ N− value [IC = initial_voltage]**

where
> The capacitor name must start with the letter **C**.
> **N+** and **N−** are the positive and negative nodes of the capacitor, respectively.
> **value** indicates the value of the capacitance.
> The initial condition for transient analysis is assigned using **IC =** initial_voltage on the capacitor.

Independent Voltage Source

The general format for describing independent voltage source is

> **Vname N+ N− [DC value] [AC magnitude phase]**
> $\quad$ **[PULSE V_1 V_2 td tr tf pw per]**
> $\quad$ **or [SIN V_O V_a freq td df phase]**
> $\quad$ **or [EXP V1 V_2 td$_1$ t$_1$ td$_2$ t$_2$]**
> $\quad$ **or [PWL t1 V_1 t$_2$ V_2 ... t$_n$,V_n]**
> $\quad$ **or [SFFM V_O V_a freq md fs]**

where
> The voltage source must start with letter **V**.
> **N+** and **N−** are the positive and negative nodes of the source, respectively.
> Sources can be assigned values for dc analysis **[DC value]**, ac analysis **[AC magnitude phase]**, and transient analysis. Only one of the transient response source options (**PULSE, SIN, EXP, PWL, SFFM**) can be selected for each source. The ac phase angle is in degrees. The transient signal generators PULSE, SIN, EXP, PWL, and SFFM are discussed in Section 1.5.

Independent Current Source

The general format for describing independent current source is

> **Iname N+ N− [DC value] [AC magnitude phase]**
> $\quad$ **[PULSE V_1 V_2 td tr tf pw per]**

> or [SIN V$_O$ V$_a$ freq td df phase]
> or [EXP V1 V$_2$ td$_1$ t$_1$ td$_2$ t$_2$]
> or [PWL t1 V$_1$ t$_2$ V$_2$ ... t$_n$,V$_n$]
> or [SFFM V$_O$ V$_a$ freq md fs]

where

The current source must start with letter **I**.

N+ and **N−** are the positive and negative nodes of the source, respectively. Current flows from a positive node to the negative node.

Independent current sources can be assigned values for dc analysis [**DC value**], ac analysis [**AC magnitude phase**], and transient analysis. Only one of the transient response source options (**PULSE, SIN, EXP, PWL, SFFM**) can be selected for each source. The ac phase angle is in degrees.

1.3   Control Statements

1.3.1   Circuit Title

The circuit title must be the first statement in the SPICE program or circuit netlist. If this is not done, the program will assume that the first statement is the circuit title. The circuit title is used to label the output when the analysis is completed. If additional comments are needed in the circuit description, the comment statement can be used.

1.3.2   Comments (*)

An asterisk, *, in the first column of a line indicates a comment line. An example of a comment is as follows:

```
*

* Comment line begins with asterisks in SPICE.
*
```

1.3.3   Operating Point (.OP)

The operating point of devices and elements in a circuit can be printed out using the **.OP** command. The general format for using the .OP control statement is

.OP

PSPICE always calculates the operating point of devices in a circuit. With **.OP** control statements, the following values are printed.

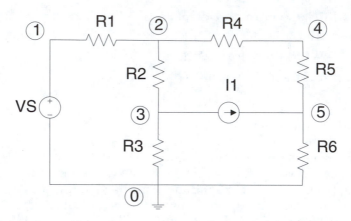

FIGURE 1.1
Resistive circuit with multiple sources.

- Voltages at each node of a circuit
- Currents and power dissipation of all voltage sources in a circuit
- Transistor diode parameters, if the previously mentioned devices are present in a circuit

Other control statements, such as **.DC** (dc analysis), **.TRAN** (transient analysis) and **.AC** (ac analysis) are discussed in the following sections. Additional control statements are covered in Chapter 2. The following example illustrates the use of .OP control statement and the element statements.

Example 1.1   Resistive Circuit with Multiple Sources

Figure 1.1 shows a resistive circuit with multiple sources. VS = 10 V, R1 = 500 Ω, R2 = 1 KΩ, R3 = 2 KΩ, R4 = 1 KΩ, R5 = 3 KΩ, R6 = 5 KΩ, and I1 = 5 mA. Find the nodal voltages.

Solution

PSPICE program:

```
Resistive Circuit with Multiple Sources
VS     1  0   DC     10V
R1     1  2   500
R2     2  3   1000
R3     3  0   2000
R4     2  4   1000
R5     4  5   3000
R6     5  0   5000
I1     3  5   DC     5mA
.OP
.END
```

The nodal voltages obtained from the PSPICE output file are given in the table below:

Node	Voltage, V
1	10.0000
2	7.9545
3	1.9697
4	9.8485
5	15.5300

1.4   DC Analysis (.DC)

The **.DC** control statement specifies the values that will be used for dc sweep or dc analysis. The general format for the **.DC** statement is

.DC SOURCE_NAME START-VALUE STOP_VALUE INCREMENT_VALUE

where
   SOURCE_NAME is the name of an independent voltage or current source.
   START_VALUE, STOP_VALUE, and **INCREMENT_VALUE** represent
      the starting, ending, and increment values of the source, respectively.

For example,

   .DC Vsource 0.5 5 .1

causes a dc source named Vsource to be swept, starting at a value of 0.5 V and stopping at 5 V, with incremental steps of 0.1 V. PSPICE analyzes the circuit at each value of Vsource.
   To perform a dc analysis with a dc source of one specified value, a **.DC** statement can be used but the start_value and stop_value are made equal to the specified value of the dc source. For example,

   .DC VCC 5 5 1

causes a dc source named VCC, with a constant voltage of 5 V, to be used for dc analysis.
   The **.DC** control statement can be used to sweep a second independent voltage source over a specified range of values. The general format for double sweep is

   .DC S1 S1_start S1_stop s1_incr S2 S2_start S2_stop S2_incr

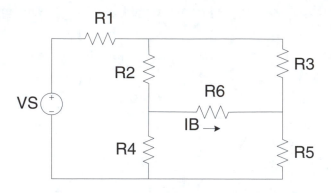

FIGURE 1.2
Bridge circuit.

where
    S1 is the name of first source. The source is swept from S1_start, and stops
        at S1_stop with increments of S1_incr.
    S2 is the name of the second source. It is swept from S2_start and sweep
        stops at S2_stop with increments of S2_incr.

The first sweep source S1 will be the "inner" loop, implying that the entire
first sweep will be performed for each value of the second sweep. The two-
source sweep is useful in generating current versus voltage characteristics
of semiconductor devices. For example,

.DC VCE 0 10V .2V IB 0mA 1mA .2mA

supplies two sources VCE and IB for sweep. VCE will vary from 0 to 10 V
with 0.2-V increments, while IB will sweep from 0 mA to 1 mA with 0.2-mA
increments. For each value of IB, VCE is swept from 0 V to 10 V. An example
that involves two sweeps can be found in Section 7.1. The following example
shows a single voltage source sweep.

Example 1.2   Bridge Circuit: Calculation of Bridge Current and DC Sweep

For the bridge circuit shown in Figure 1.2, R1 = 100 Ω, R2 = 100 Ω, R3 =
100 Ω, R4 = 400 Ω, R5 = 300 Ω, and R6 = 50 Ω. If the source voltage VS is
swept from 0 V to 10 V in increments of 2 V, find the current IB.

Solution

Figure 1.2 can be redrawn with node numbers and element names. The
redrawn circuit is shown in Figure 1.3.

PSPICE program:

```
Bridge Circuit
*
VS        1   0   DC    10V
VM        3   5   DC    0;    current monitor
R1        1   2   100
R2        2   3   100
R3        2   4   100
R4        3   0   400
R5        4   0   300
R6        4   5   50
.DC       VS  0   10    2
.PRINT        DC   I(VM)
.END
```

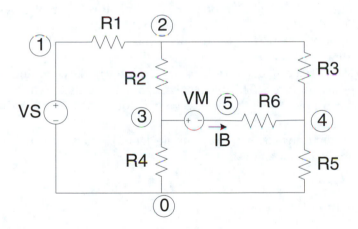

FIGURE 1.3
Figure 1.2 with node numbers and element names.

The relevant output from the program is given in the table below.

Voltage Source VS, V	Current IB, A
0.000E+00	0.000E+00
2.000E+00	3.361E−04
4.000E+00	6.723E−04
6.000E+00	1.008E−03
8.000E+00	1.345E−03
1.000E+01	1.681E−03

1.5   Transient Analysis (.TRAN)

The **.TRAN** control statement is used to perform transient analysis on a circuit. The general format of the **.TRAN** statement is

.TRAN TSTEP TSTOP <TSTART> <TMAX> <UIC>

where
   The terms inside the angle brackets are optional.
   TSTEP is the printing or plotting increment.
   TSTOP is the final time of the transient analysis.
   TSTART is the starting time for printing out the results of the analysis. If it is omitted, it is assumed to be zero. The transient analyses always start at time zero. If **TSTART** is non-zero, the transient analysis computations are done from time zero to **TSTART**, but the results are not written to output file.
   TMAX is the maximum step size that PSPICE uses for the purposes of computation. If **TMAX** is omitted, the default is the smallest value of either TSTEP or (TSTOP – TSTART)/50. **TMAX** is useful if you want the computational interval to be smaller than the TSTEP, the printing or plotting interval.
   UIC (Use Initial Conditions) is used to specify the initial conditions of capacitors and inductors. The initial conditions are specified in the element statement by adding the term IC = value for capacitors and inductors.

   Before doing transient analysis example, let us discuss the sources that can be used for transient analysis.

1.5.1   Transient Analysis Sources

There are five SPICE-supplied sources that can be used for transient analysis. They include:

   PULSE <parameters> for periodic pulse waveform
   EXP   <parameters> for exponential waveform
   PWL   <parameters> for piecewise linear waveform
   SIN   <parameters> for a sine wave
   SFFM   <parameters> for frequency-modulated waveform

   The format for specifying the above sources for transient analysis are described as follows.

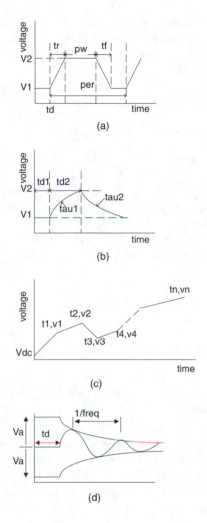

FIGURE 1.4
Transient analysis waveforms: (a) pulse waveform, (b) exponential waveform, (c) piecewise linear waveform, and (d) exponentially damped sinusoid.

Pulse Waveform

The PULSE waveform is shown in Figure 1.4(a). The general format of the pulse waveform is

PULSE(V1 V2 td tr tf pw per)

where

    V1 is the initial value of pulse. There is no default value for V1.
    V2 is the final voltage of the pulse. There is no default value for V2.
    td is delay time; its default value is zero.

tr is the rise time; its default value is the printing or plotting increment.
tf is the fall time; its default is also TSTEP.
pw is the pulse width; the default value of PW is TSTOP, the final time of
the transient analysis.
per is the period; its default is also TSTOP. The period does not include the
initial delay, td.

An example of using the PULSE statement is

VPULSE 1 0 PULSE(0V 10V 10ns 20ns 50ns 1μs 3μs)

The above statement means a signal name VPULSE is connected to nodes 1
and 0. The pulse waveform starts at 0 V and stays there for 10 ns. The voltage
increases linearly from 0 V to 10 V during the next 20 ns. The voltage stays
at 10 V for 1 μs. Then it decreases linearly from 10 V to 0 V during the next
50 ns. The cycle is repeated every 3 μs.

Exponential Waveform
An exponential waveform is shown in Figure 1.4(b). The general format of
the exponential waveform is

EXP(V1 V2 td1 tau1 td2 tau2)

where
V1 is the initial voltage in volts. V1 must be specified because it does not
have a default value.
V2 is the peak voltage in volts; it must also be specified.
td1 is the rise delay time in seconds; its default value is zero.
tau1 is the rise time constant in seconds; its default value is TSTEP, the
printing or plotting increment in .TRAN statement.
td2 is the fall delay time in seconds; its default is (td1 + TSTEP).
tau2 is the fall time constant in seconds. The default of tau2 is TSTEP.

An example of using the EXP statement is

VEXP 2 1 EXP(–1V 5V 1μs 10μs 30μs 15μs)

The above statement means the voltage VEXP, connected to nodes 2 and 1,
is an exponential waveform. The waveform is –1 V for the first 1 μs. The
voltage increases exponentially from –1 V to 5 V with a time constant of
10 μs. The voltage increase lasts for 30 μs. Then the voltage decays from 5 V
to –1 V with a time constant of 15 μs.

Piecewise Linear Waveform
The piecewise linear function is constructed using straight lines between
points. The piecewise linear waveform is shown in Figure 1.4(c). The general
format of the piecewise linear waveform is:

PWL (T1 V1 T2 V2 ... Tn Vn)

where

Each pair of time-voltage values, Tm, Vm, (where m = 1, 2, ..., n) specifies that the value of the source is Vm volts at time Tm seconds. The times are specified in increasing order. That is, $t_1 < t_2 < t_3 ... < t_n$.

An example of using the PWL statement is

VPWL 1 0 PWL (0 0 1 2 4 2 5 3 7 3 8 2 11 2 12 0)

The above statement means the voltage VPWL, connected to the nodes 1 and 0, is a piecewise linear function constructed from the following time-voltage values: (0, 0), (1, 2), (4, 2), (5, 3), (7, 3), (8, 2), (11, 2), and (12, 0).

Damped Sinusoidal Waveform
Sinusoidal source is generated using SIN. The exponentially damped sine wave is shown in Figure 1.4 (d). The general format of the sinusoid source is

SIN (Vo Va freq td df phase)

where

Vo is the offset voltage; it has no default value. It must be specified.
Va is the peak amplitude. There is no default value. It must be specified.
freq is the frequency; its default value is 1/TSTOP, where TSTOP is the final value of the transient analysis of the .TRAN statement.
td is the delay time; its default value is zero.
df is the damping coefficient; its default value is zero.
phase is the phase; its default value is zero.

The SIN function with its parameters can be used to generate the exponentially damped sine wave described by

$$v(t) = Vo + Va * \sin\left[2\pi\left(freq(t-td)\right) - \left(phase/360\right)\right]e^{-(t-td)df} \qquad (1.2)$$

An example of using the SIN statement is

VSIN 2 1 SIN(0 10 10K)

The transient sinusoidal wave VSIN is generated with zero offset voltage, an amplitude of 10 V, and a frequency of 10,000 Hz.

It should be noted that the SIN waveform is for transient analysis only. It is deactivated during ac analysis.

Frequency-Modulated Sinusoidal Function
A single frequency-modulated signal is generated using the SFFM function. The general format of the SFFM source is

SFFM(Vo Va fc mdi fs)

where
   Vo is the offset voltage. It has no default and it must be specified.
   Va is the amplitude; it must also be specified.
   fc is the carrier frequency in hertz; its default is 1/TSTOP.
   mdi is the modulation index; its default is zero.
   fs is the signal frequency in hertz.

The SFFM is described by

$$v(t) = Vo + Va * \left[\sin\left(2\pi * fc * t + mdi * \sin\left(2\pi * fs * t\right)\right) \right] \qquad (1.3)$$

An example of using SFFM function is

VINPUT 4 0 SFFM(0 5 6Meg 8 20K)

The above source produces a 6-MHz sinusoid with amplitude of 5 V modulated at 20 KHz with a modulation index of 8. The following illustrates the use of transient response source being employed for transient analysis.

Example 1.3   Transient Response of a Series RLC Circuit

For the RLC circuit shown in Figure 1.5, L = 2 H, C = 1.5 μF, and R = 1000 Ω. Find the voltage across the resistor if the input is a pulse waveform with pulse duration of 1 ms and pulse amplitude of 5 V. Assume no initial conditions.

Solution

Figure 1.6 is Figure 1.5 with node numbers and element names.

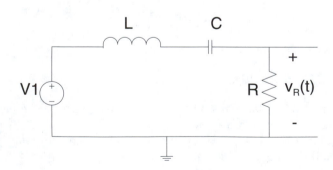

FIGURE 1.5
RLC circuit.

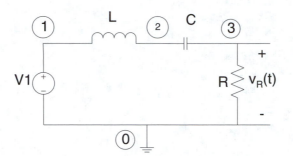

FIGURE 1.6
Figure 1.5 with node numbers and element names.

PSPICE Program:

```
RLC circuit
V1  1  0  PULSE(0  5  0.001ms  0.001ms  1ms  1)
L   1  2  2H
C   2  3  1.5e-6
R   3  0  1000
.TRAN  0.2e-3  5e-3
.PRINT  TRAN  V(3)
.PROBE
.END
```

The results are

```
TIME      V(3)
0.000E+00   0.000E+00
2.000E-04   4.711E-01
4.000E-04   8.951E-01
6.000E-04   1.267E+00
8.000E-04   1.588E+00
1.000E-03   1.858E+00
1.200E-03   2.079E+00
1.400E-03   2.253E+00
1.600E-03   2.381E+00
1.800E-03   2.468E+00
```

(continued)

```
2.000E-03    2.514E+00
2.200E-03    2.525E+00
2.400E-03    2.502E+00
2.600E-03    2.450E+00
2.800E-03    2.372E+00
3.000E-03    2.271E+00
3.200E-03    2.151E+00
3.400E-03    2.015E+00
3.600E-03    1.867E+00
3.800E-03    1.709E+00
4.000E-03    1.544E+00
4.200E-03    1.375E+00
4.400E-03    1.205E+00
4.600E-03    1.036E+00
4.800E-03    8.703E-01
5.000E-03    7.090E-01
```

1.6 AC Analysis (.AC)

The **.AC** control statement is used to perform ac analysis on a circuit. The general format of the **.AC** statement is

.AC FREQ_VAR NP FSTART FSTOP

where
> **FREQ_VAR** is one of three keywords that indicates the frequency variation by decade (DEC), by octave (OCT), or linearly (LIN).
> **NP** is the number of points; its interpretation depends on keyword (DEC, OCT, or LIN) in the FREQ. For example:
>> DEC — NP is the number of points per decade;
>> OCT — NP is the number of points per octave;
>> LIN — NP is the total number of points spaced evenly from frequency FSTART and ending at FSTOP.
>
> **FSTART** is the starting frequency. **FSTART** cannot be zero.
> **FSTOP** is the final or ending frequency.

For example, the statement

.AC LIN 100 1000 5000

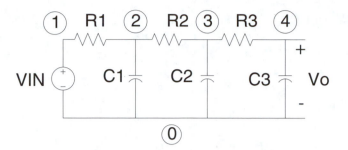

FIGURE 1.7
RC ladder network.

causes ac analysis to be performed with a frequency sweep starting at 1000 Hz and ending at 5000 Hz. The frequency range is divided into 100 equal parts; and 100 evaluations are required in this analysis.

The statement

.AC DEC 10 100 100000

causes an ac analysis to be done with frequency sweep from 100 to 100,000 Hz. There are three subintervals (100 to 1000, 1000 to 10,000, and 10,000 to 100,000). Each subinterval has ten points selected on a logarithmic scale. The following example illustrates the **.AC** command for plotting frequency response.

Example 1.4   Frequency Response of RC Ladder Network
Figure 1.7 is an RC ladder network. If R1 = R2 = R3 = 1 KΩ and C1 = C2 = C3 = 1 μF, plot the frequency (magnitude) response at the output.

Solution
The PSPICE program for obtaining the frequency (magnitude) response is

```
RC Network
VIN     1     0    AC    1   0
R1      1     2    1K
C1      2     0    1uF
R2      2     3    1K
C2      3     0    1uF
R3      3     4    1K
C3      4     0    1uF
.AC     DEC   5    10    10000
.PLOT   AC    VDB(4); plot magnitude in decibels
.PROBE ; to plot magnitude response using probe
.END
```

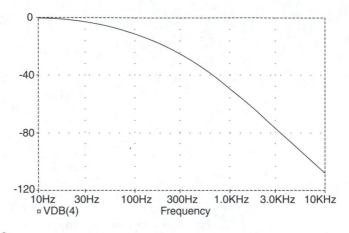

FIGURE 1.8
Frequency (magnitude) and phase of an RC ladder network.

The frequency (magnitude) response is shown in Figure 1.8.

1.7   Printing Command (.PRINT)

The **.PRINT** control statement is used to print tabular outputs. The general format of the **.PRINT** statement is

.PRINT ANALYSIS_TYPE OUTPUT_VARIABLE

where
  ANALYSIS_TYPE can be **DC, AC, TRAN, NOISE,** or **DISTO**. Only one analysis type must be specified for the **.PRINT** statement.
  OUTPUT_VARIABLE can be voltages or currents. Up to eight output variables can accompany one **.PRINT** statement. If more than eight output variables are to be printed, additional **.PRINT** statements can be used.

The output variable may be node voltages and current through voltage sources. PSPICE allows one to obtain current flowing through passive elements. The voltage output variable has the general form:

V(node 1, node 2) or V(node 1) if node 2 is node "0."

The current output variable has the general form

I (Vname)

where Vname is an independent voltage source specified in the circuit netlist.

TABLE 1.3

Name Types for AC Output Variable

Output Variable	Meaning
V or I	Magnitude of V or I
VR or IR	Real part of complex value V or I
VI or II	Imaginary part of complex number V or I
VM or IM	Magnitude of complex number V or I
VDB or IDB	Decibel value of magnitude, i.e., $20 \log_{10} |V|$ or $20 \log_{10} |I|$
VG or IG	Group delay of complex number

For PSPICE, the current output variable can also be specified as

I(Rname)

where Rname is the resistance defined in the input circuit.
For example,

.PRINT DC V(4) V(5,6) I(Vsource)

will print dc voltage at node 4, dc voltage between nodes 5 and 6, and the current flowing through an independent voltage source named Vsource. In addition, the statement

.PRINT TRAN V(1) V(7,3)

will print the voltage at node 1 and voltage between nodes 7 and 3 for a transient analysis.

For ac analysis, output voltage and current variables may be specified as magnitude, phase, real, or imaginary. Table 1.3 shows the names types for AC output variables.
For example,

.PRINT AC VDB(3) VP(3)

will plot the voltage magnitude in dB and phase in degrees of voltage at node 3.

1.8 Plotting Command (.PLOT)

The **.PLOT** control statement is used to generate line printer plots of specified output variables. The general format of the **.PLOT** command is

.PLOT ANALYSIS_TYPE OUTPUT_VARIABLE PLOT_LIMITS

where

ANALYSIS_TYPE can be **DC, AC, TRAN, NOISE,** or **DISTO.** Only one analysis type must be specified for a **.PRINT** statement.

OUTPUT_VARIABLE can be voltages or currents. The methods of specifying the output variables are similar to those in the **.PRINT** control statement described in Section 1.7.

PLOT_LIMITS specifies the lower and upper limit values that should appear on the *y*-axis for a specified output variable. The **PLOT_LIMITS** may be omitted; in that case, PSPICE assigns a plotting range of the specified output variable. The plot limits should come immediately after the output variable to which the plotting range corresponds.

For example,

.PLOT DC V(4,3) I(VIN)

will plot the dc values of the voltage between nodes 4 and 3. In addition, the current through the independent voltage source VIN will be plotted. No plotting range is specified, so PSPICE will assign a default range for plotting the *y*-axis. The range and increment of the *x*-axis should be specified in the **.DC** control statement. In addition, the statement

.PLOT AC VDB(5)(0,60)

will plot the voltage magnitude at node 5 in the range from 0 to 60 dB.

1.9   Transfer Function Command (.TF)

The **.TF** command can be used to obtain the small-signal gain, dc input resistance, and dc output resistance of a circuit by linearizing the circuit around a bias point. The format for the **.TF** command is

.TF OUTPUT_VARIABLE INPUT_SOURCE

where:

OUTPUT_VARIABLE can be voltage or current. If the **OUTPUT_VARIABLE** is a current, it is restricted to current flowing through a voltage source.

INPUT_SOURCE must be an independent voltage or current source. If the input source is a current source, then a large resistance must be connected in parallel with the current source.

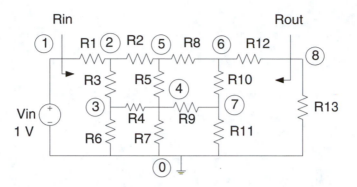

FIGURE 1.9
Resistive network.

The SPICE output file contains the following information:

- The ratio of OUTPUT_VARIABLE/INPUT_SOURCE
- The dc input resistance with respect to INPUT_SOURCE
- The dc output resistance with respect to OUTPUT_VARIABLE

The above output information is printed irrespective of the existence of **.PRINT, .PLOT,** or **.PROBE** statements in the SPICE program.
    The following example shows the application of the **.TF** command.

Example 1.5   Input and Output Resistance of a Resistive Network

A resistive network is shown in Figure 1.9. If the values of all the resistors are 10 Ω, find the input resistance R_{in} and output resistance R_{out}.

Solution

PSPICE Program:

```
RESISTIVE NETWORK
VIN   1     0     DC   1
R1    1     2     10
R2    2     5     10
R3    2     3     10
R4    3     4     10
R5    5     4     10
R6    3     0     10
R7    4     0     10
```

(continued)

R8	5	6	10
R9	4	7	10
R10	6	7	10
R11	7	0	10
R12	6	8	10
R13	8	0	10
.TF	V(8)	VIN	
.END			

PSPICE results are

```
**** SMALL-SIGNAL CHARACTERISTICS
V(8)/VIN = 6.977E-02
INPUT RESISTANCE AT VIN = 1.955E+01
OUTPUT RESISTANCE AT V(8) = 6.583E+00
```

From the results, the input dc resistance is 1.955E+01 Ohms and output dc resistance is 6.583E+00 Ohms.

The .**TF** command can be used to obtain the Thevenin equivalent circuit of a complex circuit. The Thevenin resistance can be obtained by specifying the output_variable in the .**TF** command as voltage between the specified nodes. The Thevenin voltage is obtained from the data of the node voltages obtained from the PSPICE output. The following example illustrates this application of the .**TF** command for obtaining a Thevenin equivalent circuit.

Example 1.6   Thevenin Equivalent Circuit of a Network

In Figure 1.10, R1 = 4 KΩ, R2 = 8 KΩ, R3 = 10 KΩ, R4 = 2 KΩ, R5 = 8 KΩ, R6 = 6 KΩ, and V1 = 10 V. If the current source I1 is 5 mA, find the Thevenin

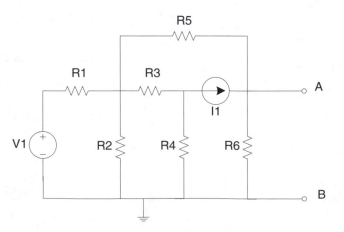

FIGURE 1.10
Circuit for Example 1.6.

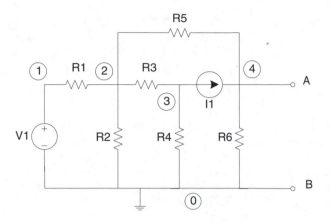

FIGURE 1.11
Figure 1.10 with node numbers.

equivalent circuit to the left of nodes A and B of the circuit. In addition, find the power dissipated in a 2-KΩ resistor that is connected between nodes A and B.

Solution

Figure 1.11 is basically Figure 1.10 with node numbers.
   The PSPICE program for obtaining the equivalent circuit is as follows.

PSPICE Program:

```
THEVENIN EQUIVALENT CIRCUIT
V1       1       0       DC 10V
R1       1       2       4K
R2       2       0       8K
R3       2       3       10K
R4       3       0       2K
R5       2       4       8K
R6       4       0       6K
I1       3       4       5MA
.TF      V(4)    V1
.END
```

The relevant data from PSPICE output is

```
NODE  VOLTAGE  NODE  VOLTAGE  NODE  VOLTAGE  NODE  VOLTAGE
(1)   10.0000  (2)    7.1910   (3)   -7.1348  (4)    20.2250
****      SMALL-SIGNAL CHARACTERISTICS
     V(4)/V1 = 2.022E-01
     INPUT RESISTANCE AT V1 = 7.574E+03
     OUTPUT RESISTANCE AT V(4)  = 3.775E+03
```

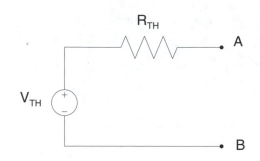

FIGURE 1.12
Thevenin equivalent circuit of Figure 1.10.

The Thevenin resistance R_{TH} is `3.775E+03` Ohms.
   The Thevenin voltage V_{TH} is the voltage at node 4, which is `20.2250` `Volts`. The Thevenin equivalent circuit is drawn in Figure 1.12.
   The power dissipated in the 2-KΩ resistor is

$$P = \left(\frac{V_{TH}}{2000 + R_{TH}} \right)^2 2000 = 0.0245 \text{ watts}$$

1.10   DC Sensitivity Analysis (.SENS)

The dc sensitivities of circuit element values and the variation of model parameters on selected output variables are obtained using the **.SENS** statement. The general format for using the **.SENS** command is

   .SENS OUTPUT_VARIABLE

where
   OUTPUT_VARIABLE can be voltage or current. If the
      OUTPUT_VARIABLE is a current, it is restricted to current flowing
      through a voltage source.

   The circuit under consideration is linearized about the bias point and the sensitivities of each output variable to all the element values and model parameters are calculated. If there are the following elements in a circuit (R_1, R_2, R_3, and V_{S1}) and the output variable of interest is V_X, then

$$V_X = f(R_1, R_2, R_3, V_{S1})$$

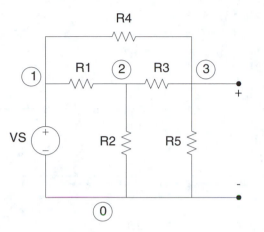

FIGURE 1.13
Bridge-T network.

The **.SENS** command will give

$$\frac{\partial V_X}{\partial R_i}, \quad \frac{\partial V_X}{\partial R_i}\left(\frac{R_i}{100}\right) \quad \text{where} \quad R_i = 1, 2, \text{ and } 3$$

$$\frac{\partial V_X}{\partial V_{S1}}, \quad \frac{\partial V_X}{\partial V_{S1}}\left(\frac{V_{S1}}{100}\right)$$

Both the absolute sensitivity $\dfrac{\partial V_X}{\partial R_i}$ or $\dfrac{\partial V_X}{\partial V_{S1}}$ and the relative sensitivity $\dfrac{\partial V_X}{\partial R_i}\left(\dfrac{R_i}{100}\right)$ or $\dfrac{\partial V_X}{\partial V_{S1}}\left(\dfrac{V_{S1}}{100}\right)$ will be outputted.

The following example illustrates the use of the **.SENS** statement.

Example 1.7   DC Sensitivity of a Bridge-T Network

In the bridge-T network shown in Figure 1.13, R1 = 20 KΩ, R2 = 40 KΩ, R3 = 20 KΩ, R4 = 50KΩ, R5 = 10 KΩ, and VS = 10 V. Use PSPICE to compute the sensitivity of the voltage across resistor R5 with respect to the circuit elements.

Solution

PSPICE Program:

```
BRIDGE-T NETWORK
VS       1   0   DC    10V
R1       1   2   20K
R2       2   0   40K
```

(continued)

```
R3        2   3   20K
R4        1   3   50K
R5        3   0   10K
.SENS    V(3)
.END
```

The relevant PSPICE results are

```
* * * * * * * * * * * * * * * * * * * * * * * * * * * * * * * * * * * * * * * * * * * * * * *

BRIDGE-T NETWORK

DC SENSITIVITY ANALYSIS  TEMPERATURE = 27.000 DEG C
* * * * * * * * * * * * * * * * * * * * * * * * * * * * * * * * * * * * * * * * * * * * * * *

DC SENSITIVITIES OF OUTPUT V(3)
ELEMENT     ELEMENT        ELEMENT           NORMALIZED
 NAME        VALUE       SENSITIVITY        SENSITIVITY
                         (VOLTS/UNIT)     (VOLTS/PERCENT)


 R1        2.000E+04     -3.289E-05         -6.578E-03
 R2        4.000E+04      8.444E-06          3.378E-03
 R3        2.000E+04     -2.400E-05         -4.800E-03
 R4        5.000E+04     -1.956E-05         -9.778E-03
 R5        1.000E+04      1.778E-04          1.778E-02
 VS        1.000E+01      2.667E-01          2.667E-02
* * * * * * * * * * * * * * * * * * * * * * * * * * * * * * * * * * * * * * * * * * * * * * *
```

The four-column tabular output has the following headers: ELEMENT
NAME, ELEMENT VALUE, ELEMENT SENSITIVITY, and NORMALIZED
SENSITIVITY. The ELEMENT SENSITIVITY is the absolute sensitivity in
amperes or volts per unit of the respective element. The NORMALIZED
SENSITIVITY is amperes or volts per 1% variation in the value of the respec-
tive element. The most informative data are the normalized sensitivities.

In the circuit, a 1% change in R1 causes about 6.6 mV variation in the
voltage at node 3. A 1% change in VS causes roughly 26.7 mV variation in
the voltage at node 3. The largest variation in the voltage at node 3 is caused
by a 1% change in VS. The smallest variation in the output voltage is caused
by a 1% change in R2. An increase in R1, R3, and R4 causes the output voltage
at node 3 to decrease; whereas an increase in R2, R5, and VS brings about
an increase in the output voltage.

1.11 Initial Conditions (.IC, UIC, .NODESET)

The **.NODESET** command is used to set the operating point at specified nodes of a circuit during the initial run of a transient analysis. The general format of the **.NODESET** command is

.NODESET V(node1) = value V(node2) = value2, ...

where
  V(node1), V(node2) are voltages at nodes 1, 2, respectively.

Voltage V(node1) is set to value1, voltage V(node2) is set to value2, and so on.
  .NODESET provides a preliminary guess for voltages at the specified nodes for bias point calculations. The **.NODESET** command is especially useful for analysis of circuits that have more than one stable state, such as bistable circuit. SPICE is guided in calculating the bias point by using the **.NODESET** command.
  The **.IC** statement is only used when the transient analysis statement, .TRAN, includes the "**UIC**" option. The initial voltage across a capacitor or the initial current flowing through an inductor can be specified as part of capacitor or inductor component statement. For example, for a capacitor we have:

Cname N+ N− value IC = initial voltage

and for an inductor, we use the statement:

Lname N+ N− value IC = initial current

It should be noted that the initial conditions on an inductor or capacitor are used provided the **.TRAN** statement includes the "**UIC**" option. The following example illustrates the use of the **.IC** command.

Example 1.8   Transient Analysis of a Sequential Circuit

For the sequential circuit shown in Figure 1.14, find the voltages across the 50-Ω resistor when the switch moved from a to b at t ≥ 0.

Solution
At t < 0, the voltage across the capacitor

$$V_C(0) = \frac{(20)*(300)}{400} = 15 \text{ V}$$

The current through the inductor is $I_L(0) = 0 \text{ A}$.

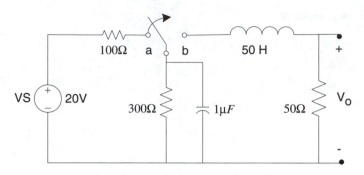

FIGURE 1.14
RLC circuit for Example 1.8.

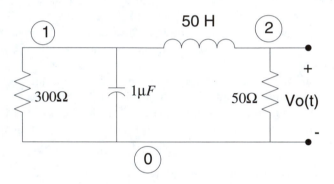

FIGURE 1.15
Equivalent circuit of Figure 1.14.

The equivalent circuit of Figure 1.14 for $t > 0$ is shown in Figure 1.15. The PSPICE program for obtaining the voltage $v_o(t)$ is

```
RLC circuit
R1      1      0      300
C1      1      0      1uF    IC = 15V
L1      1      2      50     IC = 0A
R2      2      0      50
.TRAN   0.01   0.5    UIC
.PLOT   TRAN   V(2)
.PROBE;  TO PLOT VOLTAGE AT NODE 2
.END
```

The output is plotted in Figure 1.16.

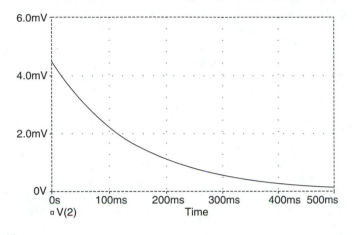

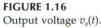

FIGURE 1.16
Output voltage $v_o(t)$.

1.12 Temperature Analysis (.TEMP)

All elements in a SPICE netlist are assumed to be measured at the nominal temperature, TNOM, of 27°C (300K). The nominal temperature of 27°C can be changed using the **OPTIONS** command. All simulations are performed at the nominal temperature. The **.TEMP** command is used to change the temperature at which a simulation is performed. The general format of the **.TEMP** statement is

   .TEMP TEMP1 TEMP2 TEMP3 ... TEMPN

where
   TEMP1, TEMP2, and **TEMP3** are temperatures at which the simulations are performed.

For example, the statement

   .TEMP 120 200

is the command to perform circuit calculations at 120°C and 200°C.
   When a temperature is changed, elements such as resistors, capacitors, and inductors have values that may change. In addition, adjustments are made to devices, such as transistors and diodes, whose models are temperature dependent.
   If one desires to sweep the circuit temperature over a range of values, the .DC sweep command can be used. The general syntax for this sweep is

.DC TEMP START_VALUE STOP_VALUE INCREMENT

where
   START_VALUE is the starting temperature in °C.
   STOP_VALUE is the ending value of temperature.
   INCREMENT is the step size.

For example, the statement

.DC TEMP 0 100 10

will make SPICE calculate all parameters of the circuit being analyzed at a starting temperature of 0°C and ending the simulation at 100°C. The increment for the analysis is 10°C.

Because a DC sweep can be nested, it is possible to sweep a component, a source, while still sweeping temperature.

1.13 PROBE Statement (.PROBE)

PROBE is a PSPICE interactive graphics processor that allows the user to display SPICE simulation results in graphical format on a computer monitor. PROBE has facilities that allow the user to access any point on a displayed graph and obtain its numerical values. In addition, PROBE has many built-in functions that enable a user to compute and display a mathematical expression that models aspects of circuit behavior.

The general format for specifying a **PROBE** statement is

.PROBE OUTPUT_VARIABLES

where
   OUTPUT_VARIABLES can be node voltages and/or devices currents. If no **OUTPUT_VARIABLE** is specified, PROBE will save all node voltages and device currents.

PSPICE creates a data file, probe.dat, for use by PROBE. The file is used by probe to display simulation results in graphical format. When PROBE is invoked, there are several commands available in the main menu for file accessing, plotting, editing, viewing, and adding or removing a trace. The PSPICE reference manual should be consulted for details.

PSPICE has several functions that PROBE can use to determine various characteristics of a circuit from variables available in PROBE. Table 1.4 shows the valid functions for PROBE expression.

TABLE 1.4

Valid Functions for PROBE Expression

Function	Meaning	Example
+	Addition of current or voltage	`V(3)+V(2,1)+V(8)`
−	Subtraction of current or voltage	`I(VS4) - I(VM3)`
*	Multiplication of current or voltage	`V(11) * V(12)`
/	Division of current or voltage	`V(6)/V(7)`
ABS(X)	$|X|$, Absolute value of X	`ABS(V(9))`
SGN(X)	+1 if X > 0; 0 if X = 0; −1 if X < 0	`SGN(V(4))`
SQRT(X)	$X^{1/2}$, square root of X	`SQRT(I(VM1))`
EXP(X)	e^X	`EXP(V(5,4))`
LOG(X)	ln(X), log base e of X	`LOG(V(9))`
LOG10(X)	$\log_{10}(X)$, log base 10 of X	`LOG10(V(10))`
DB(X)	$20*\log_{10}(X)$, magnitude in decibels	`DB(V(6))`
PWR(X,Y)	$|X|^Y$, X to the power Y	`PWR(V(2), 3)`
SIN(X)	sin(X), X in radians	`SIN(6.28*V(2))`
COS(X)	cos(X), X in radians	`COS(6.28*V(3))`
TAN(X)	tan(X), (X in radians	`TAN(6.28*V(4))`
ARCTAN(X)	$\tan^{-1}(X)$, X in radians	`ARCTAN(6.28*V(2))`
ATAN	$\tan^{-1}(X)$ (results in radians)	`ATAN (V(9)/V(4))`
d(X)	Derivative of X with respect to X-axis variable	`D(V(12))`
S(X)	Integral of X over the X-axis variable	`S(V(15))`
AVG(X)	Running average of X over the range of X-axis variable	`AVG (V5,3)`
*AVGX(XO,XF)	*Running average of X from X-axis value, XO, to the X-axis value, XF.	`AVG V(5,4)(2e-3,20e-3)`
RMS(x)	Running RMS average of X over the range of the X-axis variable	`RMS(VS2)`
MIN(X)	Minimum of real part of X	`MIN(VM3)`
MAX(X)	Maximum of real part of X	`MAX(VM3)`
M(X)	Magnitude of X	`M(V(5))`
P(X)	Phase of X (result in degrees)	`P(V(4))`
R(X)	Real part of X	`R(V(3))`
IMG(X)	Imaginary part of X	`IMG(V(6))`
G(X)	Group delay of X (results in seconds)	`G(V(7))`

Example 1.9   Power Calculations of an RL Circuit Using PROBE

For the RL circuit shown in Figure 1.17, $v(t) = 10\sin(200\pi t)$ volts. Use PROBE to plot the average power P_{AVE} delivered to the resistor R as a function of time.

Solution

Since the input voltage is a sinusoidal voltage, we use the sinusoidal waveform included as a function of the transient analysis sources, given as

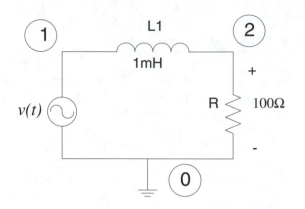

FIGURE 1.17
RL circuit.

$$sin(V_O, V_a, V_{freq}, td, df, phase)$$

In this problem,

$$V_O = 0 \text{ V} \qquad V_a = 10 \text{ V} \qquad freq = 100 \text{ Hz}$$
$$td = 0 \text{ second} \qquad df = 0 \qquad phase = 0$$

Because the analysis has time as the independent variable, transient analysis is performed.

PSPICE Program:

```
RL CIRCUIT AND PROBE
VS     1        0         SIN(0 10 100 0 0 0)
L1     1        2         1MH
R1     2        1         100
*CONTROL STATEMENTS
.TRAN  1.0E-3  3.0E-2
.PROBE
.END
```

After running the program and invoking probe, the PROBE expression that can be used to obtain the average power is

$$\text{Average Power} = rms\big(V(2) * rms(I(R1))\big)$$

The graphic display obtained from probe plot is shown in Figure 1.18.

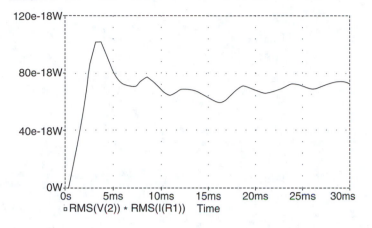

FIGURE 1.18
Average power from PROBE plot.

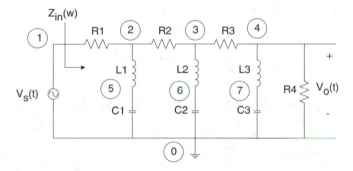

FIGURE 1.19
Passive filter network.

Example 1.10   Input Impedance vs. Frequency of a Filter Network

For the passive filter network shown in Figure 1.19, R1 = R2 = R3 = 500 Ω, R4 = 1000 Ω, C1 = C2 = C3 = 1.5 μF, L1 = 2 mH, L2 = 4 mH, and L3 = 6 mH. Use PROBE to find the input impedance $|Z_{in}(w)|$ with respect to frequency.

Solution

The PSPICE program for analyzing the circuit is

```
FILTER CIRCUIT
VS    1    0    AC    1    0
R1    1    2    500
```

(continued)

```
L1     2      5      2E-3
C1     5      0      1.5E-6
R2     2      3      500
L2     3      6      4E-3
C2     6      0      1.5E-6
R3     3      4      500
L3     4      7      6E-3
C3     7      0      1.5E-6
R4     4      0      1000
*  CONTROL  STATEMENTS
.AC    DEC    10     1.0E2      1.0E7
.PROBE
.END
```

The PROBE expression for obtaining the input impedance as a function of frequency is

$$Z_{in}(w) = VM(1)/IM(R1)$$

The plot of the input impedance is shown in Figure 1.20.

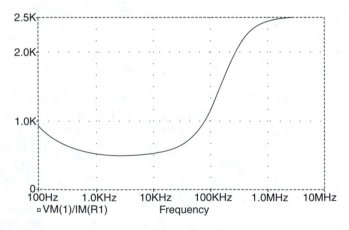

FIGURE 1.20
Input impedance vs. frequency.

Bibliography

1. Al-Hashimi, Bashir, *The Art of Simulation Using PSPICE, Analog, and Digital*, CRC Press, Boca Raton, FL, 1994.
2. Brown, William L. and Szeto, Andrew Y. J., Verifying Spice Results with Hand Calculations: Handling Common Discrepancies, *IEEE Trans. Education*, 37(4), 358–368, 1994.
3. Fenical, L. H., *PSPICE: A Tutorial*, Prentice-Hall, Englewood Cliffs, NJ, 1992.
4. Kielkowski, Ron M., *Inside SPICE, Overcoming the Obstacles of Circuit Simulation*, McGraw-Hill, New York, 1994.
5. Lamey, Robert, *The Illustrated Guide to PSPICE*, Delmar Publishers, Albany, NY, 1995.
6. Monssen, Franz, *PSPICE with Circuit Analysis*, MacMillan, New York, 1992.
7. Nilsson, James W. and Riedel, Susan A., *Introduction to PSPICE*, Addison-Wesley, Reading, MA, 1993.
8. OrCAD PSPICE A/D, Users Guide, November 1998.
9. Prigozy, Stephen, Novel Applications of PSPICE in Engineering, *IEEE Trans. Education*, 32(1), 35–38, 1989.
10. Rashid, Mohammad H., *SPICE for Power Electronics and Electric Power*, Prentice-Hall, Englewood Cliffs, NJ, 1993.
11. Rashid, Mohammad H., *SPICE for Circuits and Electronics Using PSPICE*, Prentice-Hall, Englewood Cliffs, NJ, 1990.
12. Roberts, Gordon W. and Sedra, Adel S., *Spice for Microelectronic Circuits*, Saunders College Publishing, Fort Worth, TX, 1992.
13. Thorpe, Thomas W., *Computerized Circuit Analysis with Spice*, John Wiley & Sons, New York, 1991.
14. Tuinenga, Paul W., *SPICE, A Guide to Circuit Simulations and Analysis Using PSPICE*, Prentice-Hall, Englewood Cliffs, NJ, 1988.
15. Vladimirescu, Andrei, *The SPICE Book*, John Wiley & Sons, New York, 1994.

Problems

1.1 For Figure 1.2, R1 = R2 = R3 = 100 Ω, R4 = 400 Ω, and R5 = 500 Ω. Determine the current IB if the source voltage VS = 10 V.

1.2 For Figure P1.2, L = 2 H and R = 400 Ω. If $V_s(t) = 10 \exp(-2t)\cos(1000\,\pi t)$ with a duration of 2 ms, find the output waveform $V_o(t)$.

1.3 Plot the magnitude response of the Wein-Bridge circuit shown in Figure P1.3. Assume that C1 = C2 = 4 nF, R1 = R3 = R4 = 5 KΩ and R2 = 10 KΩ. (a) What is the center frequency? (b) What is the bandwidth?

1.4 The simplified equivalent circuit of an amplifier is shown in Figure P1.4. Use SPICE to obtain the input and output resistance. What is the voltage gain at dc? Assume that RGS = 100 KΩ, Rds = 50 KΩ, RS = 50 Ω, RL = 10 KΩ, and RLC = 5 KΩ.

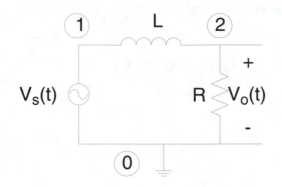

FIGURE P1.2
RL circuit.

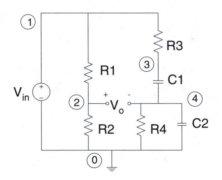

FIGURE P1.3
Wein-Bridge circuit.

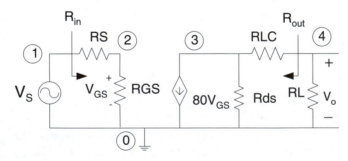

FIGURE P1.4
Simplified equivalent circuit of an amplifier.

1.5 For the multi-source resistive circuit shown in Figure P1.5, $I1 = 2$ mA, $V1 = 5$ V, $V2 = 4$ V, $R1 = 1$ KΩ, $R2 = 4$ KΩ, $R3 = 2$ KΩ, $R4 = 10$ KΩ, $R5 = 8$ KΩ, $R6 = 7$ KΩ, and $R7 = 4$ KΩ. Find the Thevenin equivalent circuit at nodes A and B.

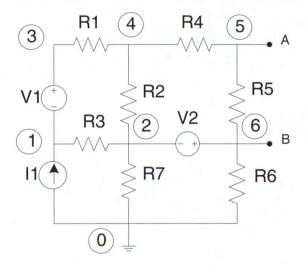

FIGURE P1.5
Multi-source resistive circuit.

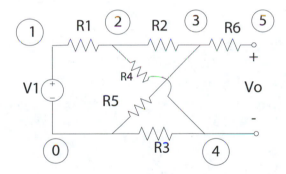

FIGURE P1.6
Resistive circuit.

1.6 Compute the sensitivity of the output voltage with respect to the circuit elements. Assume that V1 = 10 V, R2 = R3 = 4 KΩ, R4 = R5 = 8 KΩ, and R1 = R6 = 2 KΩ.

1.7 For the RLC circuit shown in Figure P1.7, V1 = 8 V, R1 = 100 Ω, R2 = 400 Ω, L1 = 5 mH, and C1 = 20 μF. The switch moves from point A to B at t = 0. Find the circuit $i(t)$ after t > 0.

1.8 For the multi-stage RC circuit shown in Figure P1.8, C1 = C2 = C3 = 1 μF, and R1 = R2 = R3 = 1 KΩ. If the input signal $V_s(t) = 25 \cos(120 \pi t + 30°)$ V, (a) determine the voltage $V_o(t)$ and (b) find the average power dissipated by R3.

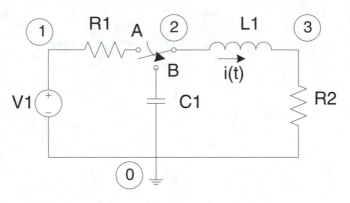

FIGURE P1.7
RLC circuit.

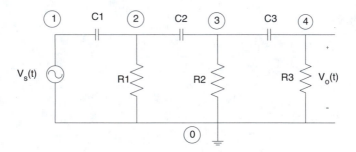

FIGURE P1.8
Multi-stage RC network.

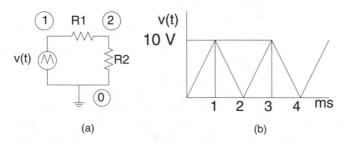

FIGURE P1.9
(a) Resistive circuit and (b) input waveform.

1.9 For the circuit shown in Figure P1.9(a), R1 = 300 Ω and R2 = 200 Ω. The input signal is a triangular wave shown in Figure P1.9(b). Use PROBE to plot the instantaneous voltage and the rms voltage across the 200-Ω resistor. What is the ratio of the rms voltage to the average voltage of the triangular wave?

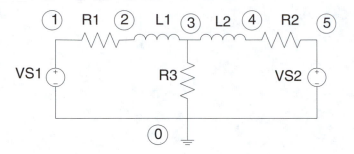

FIGURE P1.10
Circuit with two sources.

1.10 The circuit shown in Figure P1.10 has two sources with the same amplitude and frequency but different phases. R1 = R2 = 100 Ω, R3 = 5 KΩ, L1 = L2 = 1 mH, $VS1(t) = 168 \sin(120 \pi t)$ V and $VS2(t) = 168 \sin(120 \pi t + 60°)$ V. Use PROBE to determine the average power supplied to or obtained from the sources VS1 and VS2. In addition, determine the power supplied to resistor R3.

1.11 For the twin-T network shown in Figure P1.11, R1 = R4 = 1 KΩ, R2 = R3 = 2 KΩ, R5 = 5 KΩ, C1 = C2 = 1 μF, and C3 = 0.5 μF. Use PROBE to plot the magnitude of the input impedance Z_{in} with respect to frequency of source VS.

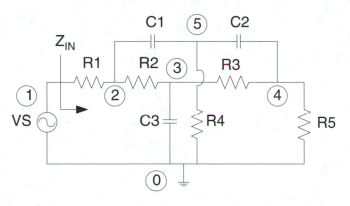

FIGURE P1.11
Twin-T circuit.

2

PSPICE Advanced Features

This chapter is a continuation of the discussion of the PSPICE features. Several of the PSPICE control statements not discussed in Chapter 1 are covered in this chapter. We briefly discuss the device models. This is followed by methodologies for changing component values. Subcircuit is defined and subcircuit calls are discussed. The analog behavior model and Monte Carlo analysis are also presented.

2.1 Device Model

The **.MODEL** statement specifies a set of device parameters which can be referenced by elements or devices in a circuit. The general form of the **.MODEL** statement is

.MODEL MODEL_NAME MODEL_TYPE PARAMETER_NAME=VALUE

where

MODEL_NAME is a name devices use to reference a particular model. The model name must start with a letter. To avoid confusion, it is advisable to make the first character of the model_name identical with the first character of the device name. See Table 1.1 for a list of the element names.

MODEL_TYPE refers to the device type, which can be active or passive. The MODEL_TYPEs available in PSPICE are shown in Table 2.1. The reference model may be available in the main circuit file, or accessed through an .INC statement, or may be in a library file. A device cannot reference a model statement that does not correspond to that type of model. It is possible to have more than one model of the same type in the circuit file but they must have different model names.

PARAMETER_NAME=VALUE follow the model type. The model parameter values are enclosed in parenthesis. It is not required to list all the parameter values of the device. Parameters not specified are assigned default values.

TABLE 2.1

Model Types of Devices

Type of Device	Model Type	Recommended Instance Name
Capacitor	CAP	CXXX
Inductor	IND	LXXX
Resistor	RES	RXXX
Diode	D	DXXX
NPN bipolar transistor	NPN	QXXX
PNP bipolar transistor	PNP	QXXX
Lateral PNP bipolar transistor	LPNP	QXXX
N-channel junction FET	NJF	JXXX
P-channel junction FET	PJF	JXXX
N-channel MOSFET	NMOS	MXXX
P-channel MOSFET	PMOS	MXXX
N-channel GaAs MESFET	GASFET	BXXX
Nonlinear, magnetic curve (transformer)	CORE	KXXX
Voltage-controlled switch	VSWITCH	SXXX
Current-controlled switch	ISWITCH	WXXX

In general, the **.MODEL** statement should adhere to the following rules:

1. More than one **.MODEL** statement can appear in a circuit file and each .MODEL statement should have a different model name. For example, the following models of MOSFETs are valid.

   M1 1 2 0 0 MOD1 L=10U W=20U
   M2 1 2 0 0 MOD2 L=10U W=40U
   .MODEL MOD1 NMOS (VTO=1.5 KP=400)
   .MODEL MOD2 NMOS (VTO=1.5KP=300)

   There are two model names, MOD1 and MOD2, for model type NMOS.

2. More than one device of the same type may reference a given model using the .MODEL statement. For example,

   D1 1 2 DMOD
   D2 2 3 DMOD
   .MODEL DMOD D (IS=1.0E-14 CJP=0.3P VJ=0.5)

   Two diodes, D1 and D2, reference the given diode model DMOD.

3. A device cannot reference a model statement that does not correspond to the device. For example, the following statements are incorrect.

R1 1 2 DMOD

.MODEL DMOD D (IS=1.0E-6)

Resistor R1 *cannot* reference a diode model DMOD.

Q1 3 2 1 MMDEL

.MODEL MMDEL NMOS (VTO=1.2)

The bipolar transistor *cannot* reference the NMOS transistor.

4. A device cannot reference more than one model in a netlist.

In the following sections, we discuss the **.MODEL** statements for both passive (R, L, C) and active (D, M, Q) elements.

2.1.1 Resistor Models

In Section 1.2, the basic description of passive elements was expressed in terms of element name, nodal connections, and component value. To model a resistor, two statements are required. The general format is

Rname NODE1 NODE2 MODEL_NAME R_VALUE

.MODEL MODEL_NAME RES[MODEL_PARAMETERS]

where

MODEL_NAME is a name preferably starting with character R. It can be up to eight characters long.

RES is the specification for PSPICE model type associated with resistors.

MODEL_PARAMETERS are parameters that can vary. Table 2.2 shows the model parameters for resistors and their default values.

PSPICE uses the model parameters to calculate the resistance using the following equations:

$$R\_value * R\left[1 + TC1(T - T_{nom}) + TC2(T - T_{nom})^2\right] \tag{2.1}$$

TABLE 2.2

Resistor Model Parameters and Their Default Values

Model Parameter	Description	Default Value	Unit
R	Resistance multiplier	1	
TC1	Linear temperature coefficient	0	°C^{-1}
TC2	Quadratic temperature coefficient	0	°C^{-2}
TCE	Exponential temperature coefficient	0	%/°C

or

$$R\_value * R\left[1.01\right]^{TCE\left(T - T_{nom}\right)} \tag{2.2}$$

where
   T is the temperature at which the resistance needs to be calculated.

Equation (2.1) uses the linear and quadratic temperature coefficients of the resistor. The coefficients TC1 and TC2 are specified in the **.MODEL** statement. Equation (2.2) is used if the resistance is exponentially dependent on temperature.
For example, the statements

R1 4 3 RMOD1 5K
.MODEL RMOD1 RES(R=1, TC1=0.00010)

describe a resistor with model name RMOD1. The **.MODEL** statement specifies that the resistor R1 has a linear temperature coefficient of +100 ppm/°C.
The statements

R2 5 4 RMOD2 10K
.MODEL RMOD2 RES(R=2, TCE=0.0010)

describe a resistor whose value with respect to temperature is given by the expression

$$R2\left(T\right) = 10,000 * 2 * \left(1.01\right)^{0.001\left(T - T_{nom}\right)} \tag{2.3}$$

where
   T_{nom} is the normal temperature that can be set with Tnom option.

2.1.2 Capacitor Models

Whereas resistors were modeled to be temperature dependent, capacitors can be both temperature and voltage dependent. In addition, capacitors can have initial voltage impressed on them. The element description and the model statements have features that incorporate the above influences on capacitors. The general format for modeling capacitors is

**CNAME NODE+ NODE– MODEL_NAME VALUE
IC=INITIAL_VALUE**
.MODEL MODEL_NAME CAP MODEL_PARAMETERS

TABLE 2.3

Capacitor Model Parameters

Model Parameter	Description	Default Value	Unit
C	Capacitance multiplier	1	—
TC1	Linear temperature coefficient	0	$°C^{-1}$
TC2	Quadratic temperature coefficient	0	$°C^{-2}$
VC1	Linear voltage coefficient	0	V^{-1}
VC2	Quadratic voltage coefficient	0	V^{-2}

where

MODEL_NAME is a name (preferably starting with character C). It can be up to eight characters long.

CAP is the PSPICE specification for the model type associated with capacitors.

MODEL_PARAMETERS are parameters that can be used to describe the capacitance value with respect to changes in temperature and voltage. Table 2.3 shows the capacitor model parameters.

PSPICE uses the model parameters to calculate the capacitance at a particular temperature, T, and voltage, V, using the following expression.

$$C(V,T) = \tag{2.4}$$

$$C\_value * C\left[1 + VC1 * V + VC2 * V^2\right] * \left[1 + TC1(T - T_{nom}) + TC2(T - T_{nom})^2\right]$$

where

T_{nom} is nominal temperature set by Tnom option.

For example, the statements

CBIAS 5 0 CMODEL 20e–6 IC=3.0
.MODEL CMODEL CAP(C=1, VC1=0.0001, VC2=0.00001,
TC1=–0.000006)

describe a capacitance which is a function of both voltage (V) and temperature (T) and whose value is given as

$$C(V,T) = 20.0 * 10^{-6}\left[1 + 0.001V + 0.00001V^2\right] * \left[1 - 0.000006(T - T_{nom})\right] \tag{2.5}$$

2.1.3 Inductor Models

Inductors are current and temperature independent. They are thus modeled in a similar manner to capacitors. The general format for modeling inductor is

TABLE 2.4

Inductor Model Parameters

Model Parameters	Description	Default Value	Unit
L	Inductance multiplier	1	—
TC1	Linear temperature coefficient	0	$°C^{-1}$
TC2	Quadratic temperature coefficient	0	$°C^{-2}$
IL1	Linear current coefficient	0	A^{-1}
IL2	Quadratic current coefficient	0	A^{-2}

LNAME NODE+ NODE− MODEL_NAME VALUE IC=INITIAL VALUE

.MODEL MODEL_NAME IND MODEL_PARAMETERS

where

MODEL_NAME is a name (preferably starting with L). It can be up to eight characters long.

IND is the PSPICE specification for model types associated with inductors.

MODEL_PARAMETERS are parameters that can be used to express the inductance value as a function of temperature and current. Table 2.4 shows the inductor model parameters.

The inductance at a particular current and temperature is given by the expression

$$L(I,T) = L_{value} * L[1 + IL1 * I + IL2 * I^2][1 + TC1(T - T_{nom}) + TC2(T - T_{nom})^2] \quad (2.6)$$

where

T_{nom} is the nominal temperature set by the Tnom option.

For example, the statements

L1 6 5 LMOD 25m IC=1.5A

.MODEL LMOD IND(L=1 IL1=0.001 TC1=−0.00002)

describe an inductor of value 25 millihenries with initial current 1.5 A whose value is a function of both current (I) and temperature (T), and it is given as

$$L(I,T) = 25.0 * 10^{-3}[1 + 0.001I] * [1 - 0.00002(T - T_{nom})] \quad (2.7)$$

The following example explores the temperature effects on a notch filter.

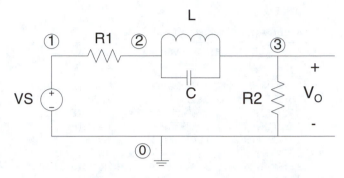

FIGURE 2.1
Notch filter.

Example 2.1   Temperature Effects on a Notch Filter

A notch filter circuit is shown in Figure 2.1, find the change in the notch frequency as the temperature increases from 25°C to 100°C. Assume the following values for TC1 and TC2:

For R, TC1 = 1.0E–5 and TC2 = 0
For C, TC1 = –6.06E–6 and TC2 = 0
For L, TC1 = 1.0E–7 and TC2 = 0

Solution

PSPICE program:

```
NOTCH FILTER AND TEMPERATURE
.OPTIONS RELTOL=1E-8
.OPTIONS NUMDGT=6
VS   1   0   AC 1   0
R1   1   2   RMOD 1K
.MODEL   RMOD   RES(R=1 TC1=1.0E-7)
L1   2   3   LMOD 10E-6
.MODEL   LMOD   IND(L=1 TC1=1.0E-7)
C1   2   3   CMOD 400E-12
.MODEL   CMOD   CAP(C=1 TC1=-6.0E-6)
R2   3   0   RMOD 1K
.AC LIN 5000 1.1E6 4E6
.TEMP   25     100
.PRINT AC VM(3)
.PROBE   V(3)
.END
```

TABLE 2.5

Magnitude Characteristics at 25°C and 100°C

Frequency, Hz	Output Voltage at 25°C, V	Output Voltage at 100°C, V
2.51200E+06	2.23325E–02	2.51167E–02
2.51258E+06	1.94193E–02	2.22063E–02
2.51316E+06	1.65048E–02	1.92944E–02
2.51374E+06	1.35893E–02	1.63811E–02
2.51432E+06	1.06730E–02	1.34668E–02
2.51490E+06	7.75638E–03	1.05519E–02
2.51548E+06	4.83962E–03	7.63651E–03
2.51606E+06	1.92304E–03	4.72104E–03
2.51664E+06	9.93068E–04	1.80576E–03
2.51722E+06	3.90840E–03	1.10903E–03
2.51780E+06	6.82266E–03	4.02304E–03
2.51838E+06	9.73556E–03	6.93597E–03
2.51896E+06	1.26468E–02	9.84751E–03
2.52012E+06	1.84631E–02	1.27574E–02
2.52070E+06	2.13676E–02	1.85710E–02

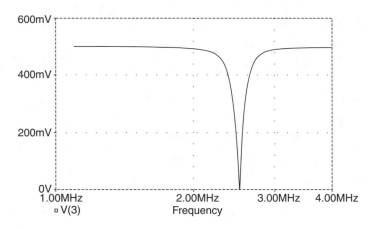

FIGURE 2.2
Notch frequency characteristic at 25°C.

Partial results for the output voltage at 25°C and 100°C are shown in Table 2.5.

From Table 2.5, the notch frequency at 25°C is 2.51664E+06 Hz and that at 100°C is 2.51722E+06 Hz. There is little shift in the notch frequency as the temperature increased from 25°C to 100°C. The magnitude characteristic at 25°C is shown in Figure 2.2.

2.1.4   Diode Models

Diode models take into account the forward and reverse bias characteristics of real diodes, the junction capacitance, the ohmic resistance of the diode,

TABLE 2.6

Diode Model Parameters

Parameters	Description	Default	Unit
IS	Saturation current	1E–14	A
N	Emission current	1	
ISR	Recombination current parameter	0	A
NR	Emission coefficient for ISR	2	
IKF	High-injection "knee" current	Infinite	A
BV	Reverse breakdown "knee" voltage	Infinite	V
IBV	Reverse breakdown "knee" current	1E–10	A
NBV	Reverse breakdown ideality factor	1	
IBVL	Low-level reverse breakdown "knee" current	0	A
NBVL	Low-level reverse breakdown ideality factor	1	
RS	Parasitic resistance	0	Ohm
TT	Transit time	0	s
CJO	Zero-bias p-n capacitance coefficient		F
VJ	p-n potential	1	V
M	p-n grading coefficient	0.5	
FC	Forward-bias depletion capacitance coefficient	0.5	
EG	Bandgap voltage (barrier height)	1.11	eV
XTI	IS temperature	3	
TIKF	IKF temperature coefficient (linear)	0	°C^{-1}
TBV1	BV temperature coefficient (linear)	0	°C^{-1}
TBV2	BV temperature coefficient (quadratic)	0	°C^{-2}
TRS1	RS temperature coefficient (linear)	0	°C^{-1}
TRS2	RS temperature coefficient (quadratic)	0	°C^{-2}
KF	Flicker noise coefficient	0	
AF	Flicker noise exponent	1	

the temperature effects on the diode characteristics, leakage current, and high-level injection of real diodes. The general format for modeling diodes is

DNAME NA NK MODEL_NAME [AREA_VALUE]
.MODEL MODEL_NAME D MODEL_PARAMETERS

where
    NA is the node number for the anode.
    NK is the node number for the cathode.
    MODEL_NAME is a name (preferably starting with D). It can be up to eight characters long.
    D is the PSPICE specification for the model associated with diodes.
    MODEL_PARAMETERS are parameters that model diode operation. Table 2.6 shows the diode model parameters.
    AREA_VALUE is used to determine the number of equivalent diodes that are connected in parallel. Parameters that are affected by the area factor include IS, CJO, IBV, and RS.

For example, the statements

D1 10 11 DMODEL
.MODEL DMODEL D(IS=1.0E–14 CJO=3PF TT=5NS BV=120V
IBV=5.0E–3)

will model diode D1 with model name DMODEL and model parameters IS, CJO, TT, BV, and IBV. The parameters not supplied will be given the default values.

2.1.5   Bipolar Junction Transistor Models

The PSPICE model of bipolar junction transistors (BJTs) is based on the integral charge-control of Gummel and Poon. For large signal analysis, the Ebers and Moll transistor model can be used. The general format for modeling bipolar transistors is

> **QNAME NC NB NE NS MODEL_NAME <AREA_VALUE>**
> **.MODEL MODEL_NAME TRANSISTOR_TYPE**
> **MODEL_PARAMETERS**

where
  NC, NB, NE, and **NS** are the node numbers for the collector, base, emitter, and substrate, respectively.
  MODEL_NAME is a name (preferably starting with the character Q). It can also be up to eight characters long.
  AREA_VALUE is the size of the transistor. It represents the number of transistors paralleled together. The parameters that are affected by the area factor include IS, IKR, RB, RE, RC, and CJE.
  TRANSISTOR_TYPE can be either NPN or PNP.
  MODEL_PARAMETERS are the parameters that model the bipolar junction transistor characteristics. Table 2.7 shows the bipolar transistor model parameters.

For example, the statements

Q1 1 3 2 2 QMOD
.MODEL QMOD NPN(IS=2.0E–14 BF=20 CJE=1.5PF CJC=200PF
TF=15NS)

describe an NPN transistor with model_name QMOD and model parameters specified for IS, BF, CJE, and TF. The undeclared parameters are assumed to have the default values.
  The statement

Q3 3 4 5 5 QMOD2
.MODEL QMOD2 PNP(IS=1.0E–15 BF=50)

describes a PNP transistor with model name QMOD2 and model parameters specified for IS and BF.

TABLE 2.7

Bipolar Transistor Model Parameters

Parameter	Meaning	Default Value	Unit
IS	p-n saturation current	1E–16	A
BF	Ideal maximum forward beta	100	
NF	Forward current emission coefficient	1	
VAF(VA)	Forward early voltage	Infinite	V
IKF(IK)	Corner for forward beta high-current roll-off	Infinite	A
ISE(C2)	Base-emitter leakage saturation current	0	A
NE	Base-emitter leakage emission coefficient	1.5	
BR	Ideal maximum reverse beta	1	
NR	Reverse current emission coefficient	1	
VAR(VB)	Reverse early voltage	Infinite	V
IKR	Corner for reverse beta high-current roll-off	Infinite	A
ISC(C4)	Base-collector leakage saturation current	0	A
NC	Base-collector leakage emission coefficient	2.0	
RB	Zero-bias (maximum) base resistance	0	Ω
RBM	Minimum base resistance	RB	Ω
IRB	Current at which RB falls halfway to RBM	Infinite	A
RE	Emitter ohmic resistance	0	Ω
RC	Collector ohmic resistance	0	Ω
CJE	Base-emitter zero-bias p-n capacitance	0	F
VJE(PE)	Base-emitter built-in potential	0.75	V
MJE(ME)	Base-emitter p-n grading factor	0.33	
CJC	Base-collector zero-bias p-n capacitance	0	F
VJC(PC)	Base-collector built-in potential	0.75	V
MJC(MC)	Base-collector p-n grading factor	0.33	
XCJC	Fraction of Cbc connected internal to RB	1	
CJS(CCS)	Collector-substrate zero-bias p-n capacitance	0	F
VJS(PS)	Collector-substrate built-in potential	0.75	V
MJS(MS)	Collector-substrate p-n grading factor	0	
FC	Forward bias depletion capacitor coefficient	0.5	
TF	Ideal forward transit time	0	s
XTF	Transient time bias dependence coefficient	0	
VTF	Transient time bias dependency on V_{bc}	Infinite	V
ITF	Transit time dependency on I_c	0	A
PTF	Excess phase at $1/(2\pi TF)$ Hz	0	°
TR	Ideal reverse transit time	0	s
EG	Bandgap voltage (barrier height)	1.11	eV
XTB	Forward and reverse beta temperature coefficient	0	
XTI(PT)	IS temperature effect exponent	3	
KF	Flicker noise coefficient	0	
AF	Flicker noise exponent	1	

2.1.6 MOSFET Models

Depending on the level of complexity, MOSFETs can be modeled by the Shichman-Hodges model, a geometry-based analytic model, a semi-empirical short-channel model, or the Berkeley short-channel IGFET model (BSIM). The general format for modeling MOSFETs is

MNAME ND NG NS NB MODEL_NAME DEVICE_PARAMETERS
.MODEL MODEL_NAME TRANSISTOR_TYPE
MODEL_PARAMETERS

where

ND, NG, NS, and **NB** are the node numbers for the drain, gate, source, and substrate, respectively.

MODEL_NAME is a name (preferably starting with the character M). It can be up to eight characters long.

DEVICE_PARAMETERS are optional parameters that can be provided for L (length), W (width), AD (drain diffusion area), AS (source diffusion area), PD (perimeter of drain diffusion), RS (perimeter of source diffusion). NRD, NRS, NRG, and NRB are the relative resistivities of the drain, source, gate and substrate in squares. M is the device "multiplier." Its default value is one. It simulates the effect of equivalent MOSFETs connected in parallel.

TRANSISTOR_TYPE can either be NMOS or PMOS.

MODEL_PARAMETERS selected depend on the MOSFET model being used. The LEVEL parameter is used to select the appropriate model. The following models are available:

LEVEL = 1 is for the Shichman-Hodges model.

LEVEL = 2 is a geometry-based analytic model.

LEVEL = 3 is a semi-empirical, short-channel model.

LEVEL = 4 is the BSIM model.

LEVEL = 5 is the SIM3 model.

Table 2.8 shows the MOSFET model parameters for levels 1, 2, and 3. Discussions on model parameters for LEVEL = 4 and LEVEL = 5 are beyond the scope of this book.

2.2   Library File

Models and subcircuits of devices and components exist in PSPICE. There are more than 5000 device models available that PSPICE users can utilize for simulation and design. The models exist in different libraries of the PSPICE package. The reader should consult the PSPICE manuals for models available for various electronic components.

The **.LIB** statement is used to reference a model or a subcircuit that exists in another file as a library. The general format for the **.LIB** command is

.LIB FILENAME.LIB

TABLE 2.8

Model Parameters of MOSFETs

Name	Model Parameter	Unit	Default	Typical
LEVEL	Model type (1,2, or 3)		1	
L	Channel length	m	DEFL	
W	Channel width	m	DEFW	
LD	Lateral diffusion (length)	m	0	
WD	Lateral diffusion (width)	m	0	
VTO	Zero-biased threshold voltage	V	0	
KP	Transconductance	A/V^2	2E–5	
GAMMA	Bulk threshold parameter	$V^{1/2}$	0	0.35
PHI	Surface potential	V	0.6	0.65
LAMBDA	Channel-length modulation (LEVEL = 1 or 2)	V^{-1}	0	0.02
RD	Drain ohmic resistance	Ω	0	10
RS	Source ohmic resistance	Ω	0	10
RG	Gate ohmic resistance	Ω	0	1
RB	Bulk ohmic resistance	Ω	0	1
RDS	Drain-source shunt resistance	Ω	∞	
RSH	Drain-source diffusion sheet resistance	Ω/square	0	20
IS	Bulk p-n saturation current	A	1E-14	1E–15
JS	Bulk p-n saturation current/area	A/m^2	0	1E–8
PB	Bulk p-n potential	V	0.8	0.75
CBD	Bulk-drain zero-bias p-n capacitance	F	0	5PF
CBS	Bulk-source zero bias p-n capacitance	F	0	2PF
CJ	Bulk p-n zero bias bottom capacitance/length	F/m^2	0	
CJSW	Bulk p-n zero-bias perimeter capacitance/length	F/m	0	
MJ	Bulk p-n bottom grading coefficient		0.5	
MJSW	Bulk p-n sidewall grading coefficient		0.33	
FC	Bulk p-n forward-bias capacitance coefficient		0.5	
CGSO	Gate-source overlap capacitance/channel width	F/m	0	
CGDO	Gate-drain overlap capacitance/channel width	F/m	0	
CGBO	Gate-bulk overlap capacitance/channel length	F/m	0	
NSUB	Substrate doping density	$1/cm^3$	0	
NSS	Surface state density	$1/cm^2$	0	
NFS	Fast surface state density	$1/cm^2$	0	
TOX	Oxide thickness	m	∞	
TPG	Gate material type: +1 = opposite of substrate; −1 = same as substrate and 0 = aluminum		1	
XJ	Metallurgical junction depth	m	0	
UO	Surface mobility	$cm^2/V{\cdot}s$	600	
UCRIT	Mobility degradation exponent (LEVEL = 2)	V/cm	1E4	
UEXP	Mobility degradation exponent (LEVEL = 2)		0	
UTRA	(not used) Mobility degradation transverse field coefficient			
VMAX	Maximum drift velocity	m/s	0	
NEFF	Channel charge coefficient		1	
XQC	Fraction of channel charge attributed to drain		1	
DELTA	Width effect on threshold		0	
THETA	Mobility modulation (LEVEL = 3)	V^{-1}	0	
ETA	Static feedback (LEVEL = 3)		0	
KAPPA	Saturation field factor (LEVEL = 3)		0.2	
KF	Flicker noise coefficient		0	1E–26
AF	Flicker noise exponent		1	1.2

where
     FILENAME.LIB is the name of the library file.

If the FILENAME.LIB is left off, then the default file is NOM.LIB. The latter library, depending on the version of PSPICE you are running, will contain devices or names of individual libraries.

One can also set up one's own library file using the file extension .LIB. The device models and subcircuits can be placed in the file. One can access the individually created libraries the same way the PSPICE supplied libraries are accessed; that is, by using

     .LIB FILENAME.LIB

One should be careful not to give the individually created library file the same name as the ones supplied by PSPICE. The following example explores the use of models in a diode circuit.

Example 2.2   Precision Diode Rectifier Characteristics

For the precision diode rectifier, shown in Figure 2.3, VCC = 10 V, VEE = −10 V, X1 and X2 are UA741 op amps, and D1 and D2 are D1N4009 diodes. If the input voltage is the triangular wave shown in Figure 2.4 with period of 2 ms, peak-to-peak value of 10 V, and zero average value, find the output voltage.

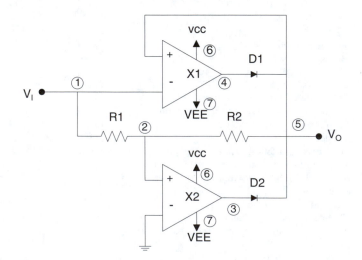

FIGURE 2.3
Precision full-wave rectifier.

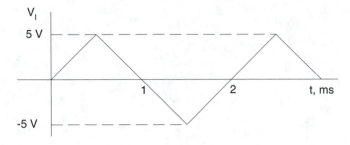

FIGURE 2.4
Input voltage.

Solution

PSPICE program:

```
FULL WAVE RECTIFIER
VIN 1   0   PWL(0 00.5M5 1.5M -5 2.5M 5 3.0M 0)
VCC 6   0   DC 10V
VEE 7   0   DC -10V
R1  1   2   10K
R2  2   5   10K
X1  1   5   6   7   4   UA741
* +INPUT; -INPUT; +VCC; -VEE; OUTPUT; CONNECTIONS FOR OP
AMP UA741
D1  4   5   D1N4009; DIODE MODEL IS DIN4009
.MODEL D1N4009 D(IS=0.1P RS=4 CJO=2P TT=3N BV=60
IBV=0.1P)
X2  0   2   6   7   3   UA741
* +INPUT; -INPUT; +VCC; -VEE; OUTPUT; CONNECTIONS FOR OP
AMP UA741
D2  3   5   D1N4009
.TRAN   0.02MS   3MS
.PROBE
.LIB NOM.LIB;
* UA741 OP AMP MODEL IN PSPICE LIBRARY FILE NOM.LIB
.END
```

The PSPICE output is shown in Figure 2.5.

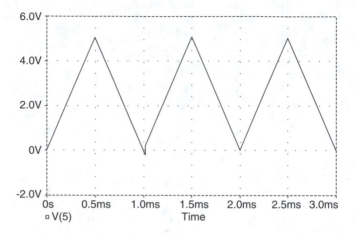

FIGURE 2.5
Output characteristics of precision full-wave rectifier.

2.3   Component Values (.PARAM, .STEP)

2.3.1   The .PARAM Statement

The **.PARAM** statement allows one to set component values using mathematical expressions. The general format for the **.PARAM** statement is

.PARAM PARAMETER_VALUE = VALUE

or

.PARAM PARAMETER_NAME = {MATHEMATICAL_EXPRESSION}

where
   PARAMETER_NAME is a set of characters allowed by PSPICE.
   PARAMETER_VALUE may be a constant or a mathematical expression. The intrinsic functions that can be used to form mathematical expressions are shown in Table 2.9.

   For example, the statement

   .PARAM C1=1.0UF, VCC=10V, VSS=–10V

defines the values of C1 as 1.0 μF, VCC = 10 V, and VSS = –10 V.
   For the following two statements:

   .PARAM RA = 10K
   .PARAM RB = {5*RA}

the first statement sets RA to be 10 KΩ and the second statement equates RB to be 5 times RA.

TABLE 2.9

Valid Functions for Mathematical Expressions

Function	Meaning	Comments
ABS(X)	$\lvert X \rvert$	Absolute value of X
ACOS(X)	$\cos^{-1}(X)$	X is between −1 and 1
ARCTAN(X)	$\tan^{-1}(X)$	Results in radians
ASIN(X)	$\sin^{-1}(X)$	$-1.0 \leq X \leq 1.0$
ATAN(X)	$\tan^{-1}(X)$	Results in radians
ATAN2(Y,X)	$\tan^{-1}(Y/X)$	Results in radians
COS(X)	$\cos(X)$	X in radians
DDT(X)	Time derivative of X	
EXP(X)	e^X	
IMG(X)	Imaginary part of X	Returns 0.0 for real number
LOG(X)	$\ln(X)$	Log base e of X
LOG10(X)	$\log(X)$	Log base 10 of X
MAX(X,Y)	Maximum of X and Y	
MIN(X,Y)	Minimum of X and Y	
M(X)	$\lvert X \rvert$	Magnitude of X, same as ABS(X)
P(X)	Phase of X	Returns 0.00 for real numbers
PWR(X,Y)	$\lvert X \rvert^Y$	Absolute value of X raised to the power Y
R(X)	Real part of X	
SDX(X)	Time integral of X	Only used for transient analysis
SGN(X)	Signum function	
SIN(X)	$\sin(X)$	X is in radians
SQRT(X)	$X^{1/2}$	Square root of X
TAN(X)	$\tan(X)$	X is in radians

The following points should be observed while using the parameter definition statement **.PARAM**.

1. If the parameter is defined by an expression, then the curly brackets { } are required.

2. **.PARAM** can be used inside a subcircuit definition to create parameters that are local to the subcircuit.

3. There are PSPICE predefined parameters, such as TEMP, VT, GMIN, and TIME. Parameter names defined with the **.PARAM** statement should be different in name from the PSPICE predefined parameters.

4. Once defined, a parameter can be used in place of all numeric values in a circuit description. For example:

    .PARAM TWO_PI = {2.0*3.14159}, F0 = 5KHZ
    .PARAM FREQ = {TWO_PI *F0}

5. **.PARAM** statements can be in a library. If the PSPICE simulator does not find parameters in a circuit file, it will search libraries for the parameters.

2.3.2   The .STEP Function

The **.STEP** function can be used to vary a circuit element signal source or a temperature over a specified range. This feature of PSPICE allows the user to observe like effects of changing a circuit element on the response of the circuit. The general formal for using the statement is

.STEP SWEEP_TYPE SWEEP_NAME START_VALUE END_VALUE INCNP

or

.STEP SWEEP_NAME LIST <VALUES>

where
> **SWEEP_TYPE** can be LIN, OCT, or DEC.
>> For LIN, we have linear sweep. The sweep variable is swept linearly with the **INCNP** being the step size from the starting value to ending value.
>>
>> For OCT, (sweep by octave), the sweep variable is swept logarithmically by octave from the start value to the ending value. However, **INCNP** is now the number of steps per octave.
>>
>> For DEC, (sweep by decade), the sweep variable is swept logarithmically by decade from the start value to ending value. **INCNP** is the number of steps per decade.
>
> **SWEEP_NAME** is the sweep variable name. It can be a model parameter, temperature, global parameter and independent voltage, or current source. During the sweep, the source's voltage or current is set to the sweep value. For example, the statements

> VCE 5 0 DC 10V
>
> .STEP VCC 0 10 2

> will cause VCC to be swept linearly from 0 to 10 V with 2-V steps.
>
> **MODEL_PARAMETER**: The model type and model name are followed by a model parameter name in parenthesis. The parameter in the model is set to the sweep value. For example, the statements

> R1 5 6 RMOD 1
>
> .MODEL RMOD RES(R = 1)
>
> .STEP RES RMOD(R) 1000 3000 500

> will cause PSPICE to analyze the circuit for the following values of R1: 100, 1500, 2000, 2500, and 3000 Ω. In the above example, RMOD is the model name, RES is the model type, and R is the parameter within the model to step. It should be noted that the

Computed value of R1 = Line value of R1 multiplied by R   (2.8)

where

Line value of R1 is the value at the end of the statement line that describes R1.

R is the value of R in the model statement.

In another example, the statements

C2 2 0 CMODEL 1.0E–9

.MODEL CMODEL CAP (C = 1)

.STEP CAP CMODEL(C) 2 22 4

will cause PSPICE to analyze the circuit with the above statements for the following values of C2: 2E–9, 6E–9, 10E–9, to 22E–9 Farads.

TEMPERATURE: Keyword **TEMP** is used for the sweep variable name. It should be followed by keyword **LIST**. The temperature is set to the sweep value. For each value in the sweep, all the circuit components have their model parameters updated to that temperature. For example, the statements

V1 1 0 DC 5

R1 1 2 1K

C1 2 0 1U

.STEP TEMP LIST 0 27 50 100

.OP

imply that for each temperature value 0, 27, 50, and 100, the component values will be updated and the operating point of the component will be determined.

GLOBAL_PARAMETER: Keyword PARAM followed by the parameter name. The latter parameter name is set to sweep. During the sweep, the global parameter's value is set to the sweep value and all expressions are evaluated. For example, the statements

VIN 1 0 AC 1

R1 1 2 100

.PARAM (CVAL = 5E–6; original value of C1)

C1 2 0 {CVAL}

.STEP PARAM CVAL 4E–6 16E–6 2E–6

* Vary C1 from 4E–6 to 16E–6 by steps of 2E–6F

.AC LIN 100 1E4 5E4

.PROBE V(2)

.END

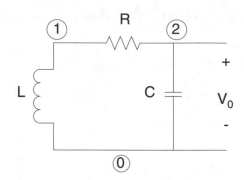

FIGURE 2.6
RLC circuit.

will obtain value of the voltage V(2) for values of C1 swept from 4 µF to 16 µF with 2-µF steps.

The following points should be noted while using the **.STEP** statement

1. The **.STEP** statement causes all analyses specified in a circuit file (**.DC, .AC, .TRAN**) to be done for each step.
2. The start value may be less than or greater than the end value of the **.STEP** statement.
3. The sweep increment value or the number of points per decade or octave (**INCNP**) should be greater than zero.

Example 2.3   Effect of Damping on an RLC Circuit

For the RLC circuit shown in Figure 2.6, R = 1 Ω, L = 1 H, and the initial voltage across capacitor is 3.3 V. If the capacitor C assumes the values of 1, 2, and 3 Farads, determine the voltage across the capacitor with respect to time.

Solution

PSPICE program:

```
THREE CASES OF DAMPING
**

L    1  0  1
R    1  2  1
C    2  0  {C1} IC=3.3V
.PARAM C1=1.0; ORIGINAL VALUE
.STEP PARAM C1 1 3 1; VARY C1 FROM 1,2,3 F
```

(continued)

```
.TRAN 0.1 10 UIC
.PLOT TRAN V(2)
.PROBE V(2)
.END
```

The plot of the three cases of damping for RLC circuit is shown in Figure 2.7.

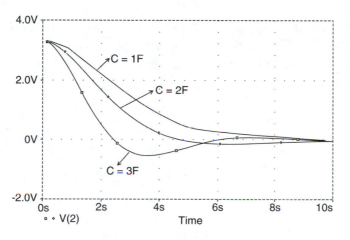
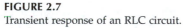

FIGURE 2.7
Transient response of an RLC circuit.

2.4   Function Definition (.FUNC, .INC)

2.4.1   The .FUNC Statement

The function statement is used to define "functions" that may be used in expressions similar to those discussed in Section 2.3.1. The functions are user defined and flexible. Because expressions are restricted to a single line, several subexpressions can be defined in a circuit file using the function statement to obtain an expression that satisfy ones application. The general form of the function statement is

   .FUNC FUNCT_NAME(ARG) {BODY}

where
   FUNCT_NAME is a name of the function with argument, arg. The
      FUNCT_NAME must be different from built-in functions, as shown in
      Table 2.9.

ARG is the argument for the function. Up to ten arguments can be used in a definition. The number of arguments in a function must agree with the number in the function definition. A function can be defined with no arguments, but the parentheses are still required.

BODY of a function definition may refer to other functions previously defined. The body of a function is enclosed is curly brackets { }.

The following points should be borne in mind when using the **.FUNC** statement.

1. The **.FUNC** statement must precede the first use of the function name **FUNCT_NAME**.
2. The body of a function definition must fit on one line.
3. **.FUNC** statements, if they appear in subcircuits, are local to those subcircuits.

If an application has several .FUNC statements, the user can create a file that contains the .FUNC definitions and access the function definitions with an **.INC** statement near the beginning of the circuit file. The next section describes the **.INC** statement. The following example describes the use of the .FUNC statement.

Example 2.4   Thermistor Characteristics

A thermistor is a device whose resistance is highly dependent on temperature. It can be used to measure temperature. The resistance, R_T of a thermistor can be expressed as

$$R_T = R_O \exp\left(\beta\left(\frac{1}{T} - \frac{1}{T_O}\right)\right) \tag{2.9}$$

where
R_T is the resistance of a thermistor at temperature T, in degrees Kelvin.
R_O is the resistance at T_O, in degrees Kelvin, which is usually taken to be 298K (≡25°C).
β is a characteristic temperature that varies with material composition of a thermistor. A typical value is 4000K, but it can vary from 1500K to 6000K.

Figure 2.8 shows a simple thermistor circuit; VS = 10 V and RS = 25 KΩ. The thermistor resistance at temperature 298K is 25 KΩ. The characteristic temperature of the thermistor is 4000K. Determine the voltage across the thermistor with respect to temperature.

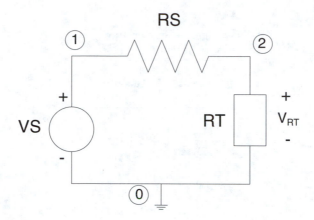

FIGURE 2.8
Thermistor circuit.

Solution

The PSPICE program for obtaining the thermistor characteristics is as follows.

PSPICE program:

```
THERMISTOR CIRCUIT
*     TO = 298
*     RO = 25000
*     B = 4000
.PARAM TS = 300
* Calculates resistance of thermistor at temperature TS
.FUNC E(X) {EXP(X)}
.FUNC RT(Y) {25000*(E(4000*((1/Y) - 0.00336)))}
.STEP PARAM  TS  300  400  10
*
VS   1   0   DC 10
R1   1   2   25K
RT   2   0   {RT(TS)}
.DC VS 10 10 1
.PRINT DC V(2)
.END
```

Table 2.10 shows the temperature vs. output voltage of the thermistor circuit.

TABLE 2.10

Voltage vs. Temperature
of a Thermistor

Temperature (K)	Voltage across Thermistor (V)
300	4.734
310	3.689
320	2.809
330	2.110
340	1.577
350	1.180
360	0.888
370	0.673
380	0.515
390	0.398
400	0.311

2.4.2   The .INC Statement

The **.INC** statement can be used to insert the content of another file into a circuit file. The general format for using the **.INC** statement is

.INC "FILENAME"

where
   FILENAME is a character string, which is a legal filename for the computer system on which the user is running the PSPICE package.

The included file may contain PSPICE valid statements with the following exceptions.

1. No title line is allowed. Instead of the title line, a comment line can be used.
2. An **.END** statement is not required. However, if an **.END** statement is present, it should mark the end of the included file.
3. An **.INC** statement can be used in the included file. However, only up to four levels of "including" are allowed.

It should be noted that including a file by using the **.INC** statement brings the file's text into the circuit file and takes up space in the main memory (RAM).

2.5   Subcircuit (.SUBCKT, .ENDS)

If a block of circuit is repetitively used in an overall circuit, then the block circuit can be defined as a subcircuit. The subcircuit can then be used repetitively. The subcircuit concept is similar to that of subroutines in programming languages, such as Fortran or C. The general form for subcircuit description is

.SUBCKT SUBCIRCUIT_NAME NODE1 NODE2 ...
[PARAMS:NAME=<VALUE>]
DEVICE STATEMENTS
.ENDS [SUBCIRCUIT_NAME]

The subcircuit definition begins with the **.SUBCKT** statement. The subcircuit definition ends with an **.ENDS** statement. The statements between **.SUBCKT** and **.ENDS** are included in the subcircuit definition.

A subcircuit definition contains only device statements. The subcircuit definition may also contain **.MODEL**, **.PARAM**, or **.FUNC** statements.

.ENDS subcircuit name indicates the end of subcircuit circuit description statements. The subcircuit_name after the **.ENDS** statement can be omitted. However, it is advisable to have the subcircuit definition that is being terminated. This is especially useful if there is more than one subcircuit that is called by the main circuit.

The symbol for a subcircuit call is **X**. It is essential that a unique name be given to each separate call of a subcircuit. The general form for a subcircuit call is

XNAME NODE1 NODE 2 ... SUBCIRCUIT_NAME
[PARAMS:NAME=VALUE]

where
   XNAME is the device name of the subcircuit. **XNAME** can be up to eight
      characters long.
   NODE1, NODE2: There must be the same number of nodes in the subcir-
      cuit calling statement as in its definition. The nodes in the subcircuit call
      must match those of the subcircuit definition.
   SUBCIRCUIT_NAME is the name of the subcircuit to be inserted into the
      main circuit (or calling circuit).

It should be noted that subcircuits can be nested. That is, subcircuit X can call other subcircuits. In addition, nesting of subcircuits cannot be circular. That is, if subcircuit X contains a call to subcircuit Y, then subcircuit Y must not contain a call to subcircuit X.

Subcircuits have the following advantages:

1. They reduce the size of circuit file, provided the circuit has repetitive elements. The repetitive parts of the circuit can be defined as
   a subcircuit and the main circuit issues subcircuit calls.
2. Once a subcircuit is defined, it can be used by other circuits. For
   example, once a subcircuit of a particular op amp is known, it can

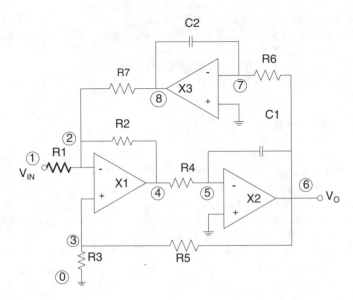

FIGURE 2.9
State-variable active filter.

be used to build and simulate amplifiers, oscillators, and filters that
use that particular op amp.

3. Subcircuits allow hierarchal testing and design of complex circuits.
   A complex circuit can be divided into the sum of parts. If some of
   the parts are repetitive, subcircuit definitions can be done for those
   repetitive parts. Once the various parts of the complex circuit are
   simulated and their functionality validated, then they can be used
   to build more complex circuits.

The following example illustrates the use of subcircuits.

Example 2.5 Frequency Response of a State-Variable Active Filter

For the state-variable active filter shown in Figure 2.9, the op amp has an
input impedance of 10^{12} Ω, an open loop gain of 10^7, and a zero output
resistance. R1 = 80 KΩ, R2 = 15 KΩ, R3 = 500 Ω, R4 = 5 KΩ, R5 = 200 KΩ,
R6 = 5 KΩ, R7 = 200 KΩ, C1 = C2 = 2 nF. Determine the frequency response
of the filter.

Solution

The op amp symbol and its simplified equivalent circuit are shown in
Figure 2.10.

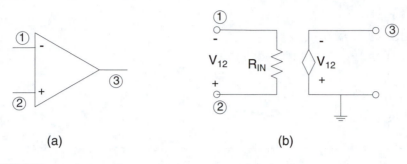

FIGURE 2.10
Op amp (a) block diagram and (b) its simplified equivalent circuit.

The subcircuit definition is

```
.SUBCKT OPAMP 1 2 3
* - INPUT; + INPUT; OUTPUT
RIN 1 2 1.0E12
EVO 0 3 1 2 1.0E7
.ENDS OPAMP
```

PSPICE program:

```
STATE-VARIABLE
VIN 1   0   AC 1   0
R1   1   2   80K
X1   2   3   4   OPAMP
R2   2   4   15K
R3   3   0   500
R4   4   5   5K
X2   5   0   6   OPAMP
C1   5   6   2nF
R5   3   6   200K
R6   6   7   5K
X3   7   0   8   OPAMP
C2   7   8   2nF
R7   8   2   200K
.AC DEC10 1E21E6
.PRINT AC VM(6)
```

(continued)

```
.PROBE
.SUBCKT OPAMP 1 2 3
*  -  INPUT;  +  INPUT;  OUTPUT
RIN 1   2   1E12
EVO 0   3   1   2   1.0E7
.ENDS OPAMP
.END
```

The magnitude response of the filter is shown in Figure 2.11.

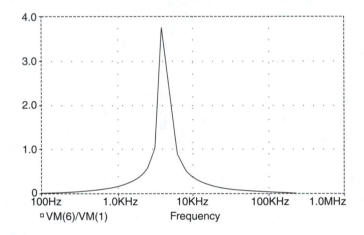

FIGURE 2.11
Frequency (magnitude) response of a state-variable filter.

2.6   Analog Behavioral Model

One way of simulating circuits is to describe the circuit in terms of the components and the connection between the components. The latter can be resistors, capacitors, inductors, transistors, voltage, and current sources. This type of simulation is called structure or primitive-level simulation. Most of the PSPICE simulations performed thus far have been of the primitive-level simulation type. If the circuit contains a lot of elements, the primitive-level simulation requires a long simulation time.

When a designer is interested in system performance, primitive-level simulation of all the subsystems may be overly detailed, time-consuming, and ineffective. At the system level, a block-diagram simulation approach might be desirable and appropriate. The function of the blocks representing a system can be described by their mathematical behavior, expressions, or relations. This type of simulation is called *analog behavioral modeling simulation*. This section discusses the analog behavioral model simulation approach of PSPICE.

The general format for using the analog behavioral model (ABM) is

ENAME CONNECTING_NODES ABM_KEYWORD ABM_FUNCTION

or

GNAME CONNECTING_NODES ABM_KEYWORD ABM_FUNCTION

where

ENAME or **GNAME** is the component name assigned to the E or G device.

CONNECTING_NODES specify the "+node" and "−node" between which the component is connected.

ABM_KEYWORD specifies the form of transfer function to be used. One of the following functions can be used.

- VALUE: Arithmetic expression
- TABLE: Lookup table
- FREQ: Frequency response
- LAPLACE: Laplace transform

ABM_FUNCTION specifies the transfer function as a mathematical expression, lookup table, or ratio of two polynomials.

The two controlled sources that can be used for the ABM are voltage-controlled voltage-source or voltage controlled current source. Thus, for modeling voltage sources, the PSPICE component should start with the letter **E**. Similarly, for modeling current sources, the PSPICE component should start with the letter **G**. The following subsection describes the analog behavioral model functions.

2.6.1   The VALUE Extension

The **VALUE** extension allows transfer functions to be written as mathematical expressions. The general forms are

ENAME N+ N− VALUE = {(EXPRESSION)}

or

GNAME N+ N− VALUE = {(EXPRESSION)}

where

(EXPRESSION) is a mathematical expression that may contain arithmetic operators (+, –, *, /), the PSPICE built-in function shown in Table 2.9, constants, node voltages, currents, and the parameter TIME. The last variable is a PSPICE interval sweep variable employed in transient analysis.

It should be noted that:

1. **VALUE** in the statement line should be followed by a space.
2. **(EXPRESSION)** must fit on a single line. If it cannot, start the following line by + and continue writing the value of the expression.

Some valid **VALUE** statements are

EAVE   1 0 VALUE = {.25*V(2,0)+(V(2,0)+V(2,3)+V(3,0)}

GVMW 4 0 VALUE = {10*cos(6.28*TIME)}

The following example illustrates the use of the **VALUE** extension.

Example 2.6   Voltage Multiplier

A voltage multiplier (Figure 2.12) has an output given as

$$V_0 = k\left[V_1(t) * V_2(t)\right] \tag{2.10}$$

If $V_1(t)$ and $V_2(t)$ are triangular and sinusoidal waveforms, find the output voltage. Assume that $k = 0.4$ and $R_O = 100\ \Omega$.

Solution
PSPICE program:

```
VOLTAGE MULTIPLIER
V1   1   0   PWL(0 0 1MS 5V 3MS -5V 5MS 5V 6MS 0)
.PARAM K = 0.4
V2   2   0   SIN(0 5 250 0 0 0)
* MULTIPLIER MODEL
EMULPLY 3 0 VALUE = {K*V(1,0)*V(2,0)}
RO   3   0   100
.TRAN 0.02MS 6MS; TRANSIENT RESPONSE
.PROBE
.END
```

The output voltage is shown in Figure 2.13.

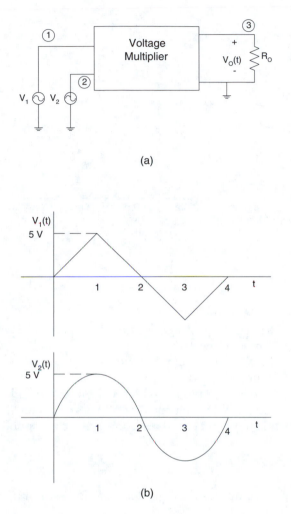

(a)

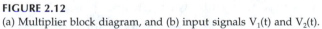

(b)

FIGURE 2.12
(a) Multiplier block diagram, and (b) input signals $V_1(t)$ and $V_2(t)$.

2.6.2 The TABLE Extension

The **TABLE** extension can be used to describe the operation of a circuit or a device by a lookup table. The general form of the **TABLE** function is

   ENAME N+ N− TABLE{EXPRESSION} = <<INPUT VALUE>
   <OUTPUT VALUE>>

or

   GNAME N+ N− TABLE{EXPRESSION} = <<INPUT VALUE>
   <OUTPUT VALUE>>

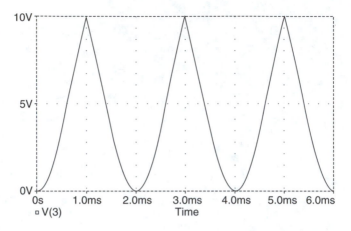

FIGURE 2.13
Output of a voltage multiplier.

where
  N+, N− are the positive and negative nodes, respectively, between which the component is connected.
  TABLE is the keyword showing that the controlled sources are described by tabular data.
  EXPRESSION is the input to the table, which is evaluated based on the tabular data. The table itself consists of pairs of values. The first value in each pair is the input and the second is the corresponding output.

It should be noted that:

1. The table's input must be in order from lowest to highest.
2. Linear interpolation is performed between entries.
3. The **TABLE** extension can be used to describe circuits or devices that can be represented by measured data.
4. The **TABLE** keyword must be followed by a space.
5. The input to the table is <expression> that must fit on one line.

The following example uses the **TABLE** expression to find a diode current.

Example 2.7   Current in a Diode Circuit
In Figure 2.14, R1 = 5 KΩ, R2 = 5 KΩ, R3 = 10 KΩ, and R4 = R5 = 10 KΩ. Table 2.11 describes the current-voltage characteristics of the diode. Find the current flowing through the diode.

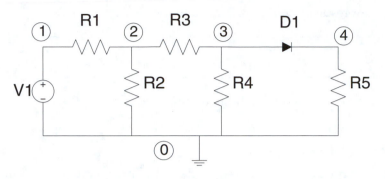

FIGURE 2.14
A diode circuit.

TABLE 2.11

Diode Characteristics

Forward Voltage (V)	Forward Current (A)
0	0
0.1	0.13E–11
0.2	1.8E–11
0.3	24.1E–11
0.4	0.31E–8
0.5	4.31E–8
0.6	58.7E–8
0.7	7.8E–6

Solution

PSPICE program:

```
DIODE CIRCUIT
V1   1   0   DC  15V
R1   1   2   5K
R2   2   0   5K
R3   2   3   10K
R4   3   0   10K
GDIODE3 4   TABLE  {V(3,4)}=(0 0) (0.1 0.13E-11) (0.2 1.8E-11)
+  (0.3 24.1E-11) (0.4 0.31E-8) (0.5 4.31E-8) (.6 58.7E-8)
+(0.7 7.8E-6)
R5   4   0   10K
.DC  V1  15  15  1
.PRINT DC I(R5)
.END
```

Relevant results from PSPICE simulation are

```
    V1            I(R5)
1.500E+01   7.800E-06
```

Thus, the current flowing through the diode is a 7.8E–06 A.

2.6.3   The FREQ Extension

The **FREQ** function can be used to describe the operation of a circuit or system by a frequency response table. The general form of the **FREQ** function is

ENAME N+ N– FREQ{EXPRESSION} = FREQUENCY VALUE, MAGNITUDE IN DB, PHASE VALUE

or

GNAME N+ N– FREQ{EXPRESSION} = FREQUENCY VALUE, MAGNITUDE IN DB, PHASE VALUE

where
   N+, N– are the positive and negative nodes, respectively, between which the component is connected.
   FREQ is the keyword showing that the controlled sources are described by a table of frequency response.
   EXPRESSION is the input to the table. The table consists of frequency value and its corresponding magnitude in decibels (dB) and phase (degrees). Interpolation is done between entries. Phase is linearly interpolated and the magnitude logarithmically interpolated. The frequencies in the table must be in order from lowest to highest.

The **FREQ** and **TABLE** functions are similar in use. Both functions are described by tabular data. The **FREQ** function is used to describe a circuit or system in terms of frequency response points (frequency, magnitude, phase). However, the **TABLE** function is used to describe circuit, device, or system operation in terms of (x, y) values.
   Note that:

   1. The **FREQ** keyword must be followed by a space.
   2. The **EXPRESSION** must fit on one line.

The following example describes the application of the **FREQ** function for plotting experimental data obtained from a filter.

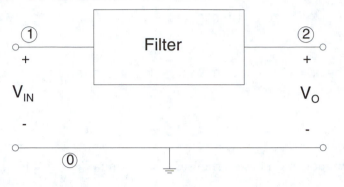

FIGURE 2.15
Block diagram of a filter.

TABLE 2.12

Frequency Response of a Filter

Frequency (KHz)	Magnitude (dB)	Phase (degrees)
1.0	−14	107
1.9	−9.6	90
2.5	−5.9	72
4.0	−3.3	55
6.3	−1.6	39
10	−0.7	26
15.8	−0.3	17
25	−0.1	11
40	−0.05	7
63	−0.02	4
100	−0.008	3

Example 2.8　Frequency Response of a Filter

A block diagram of a filter is shown in Figure 2.15. Data obtained from the filter are shown in Table 2.12. Plot the magnitude and phase responses.

Solution

PSPICE program:

```
FILTER CHARACTERISTICS
VIN 1  0  AC 1  0
R1  1  0  1K
EFILTER 2 0 FREQ {V(1,0)}=(1.0K, -14, 107) (1.9K, -9.6, 90)
+(2.5K, -5.9, 72) (4.0K, -3.3, 55) (6.3K, -1.6, 39) (10K,
-0.7, 26)
```

(continued)

```
+(15.8K, -0.3, 17) (25K, -0.1, 11) (40K, -0.05, 7) (63K,
-0.02, 4) (100K, -0.008, 3)
R2   2   0   1K
*INPUT NODES 1 AND 0, AND OUTPUT IS BETWEEN NODES 2 AND 0.
.AC DEC 5 1000 1.0E5
.PROBE V(2) V(1)
.END
```

The magnitude and phase responses are shown in Figures 2.16(a) and (b).

2.6.4   The LAPLACE Extension

The **LAPLACE** function can be used to describe the operation of a circuit or system by means of a transfer function given in terms of a Laplace transform function. The general form of the **LAPLACE** function is

ENAME N+ N– LAPLACE{EXPRESSION} = {TRANSFORM}

or

GNAME N+ N– LAPLACE{EXPRESSION} = {TRANSFORM}

where
  N+ N– are positive and negative nodes between which the component is connected.
  LAPLACE is the keyword showing that the controlled sources are described by the Laplace transfer variable S.
  EXPRESSION is the input to the transform. It follows the same rules mentioned in Section 2.6.1. It can be a voltage, current, or mathematical expression containing voltage, current arithmetic operators, and PSPICE built-in functions.
  TRANSFORM is an expression given by a ratio of two polynomials in the Laplace variable S.

Both the **LAPLACE** and **FREQ** functions can be used to model the frequency response of a circuit or system. The LAPLACE function is appropriate if the transfer function of the circuit or system is given in terms of the Laplace transform variable S. On the other hand, the **FREQ** function is used if the transfer function of the circuit or system is given in terms of a frequency response table. The **LAPLACE** function lends itself to changing the polynomial coefficients to ascertain their effects on the circuit response. The following example describes the use of the **LAPLACE** function.
  Note that:

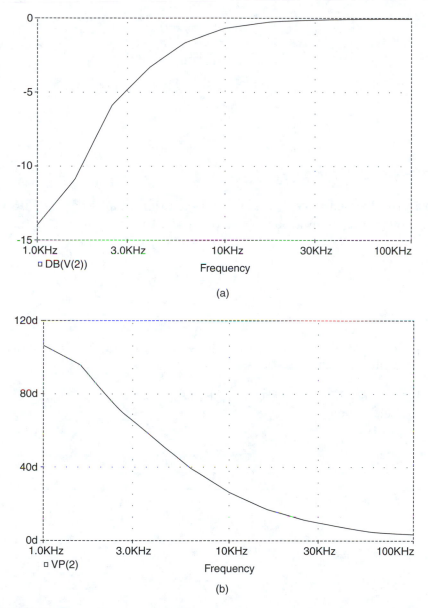

FIGURE 2.16
(a) Magnitude response and (b) phase response of a filter.

1. **LAPLACE** must be followed by a space.
2. **EXPRESSION** and **TRANSFORM** must each fit on one line.
3. Voltage, current, and **TIME** must not be used in a **LAPLACE** transfer.
4. One can use both transient and ac analysis with the **LAPLACE** function.

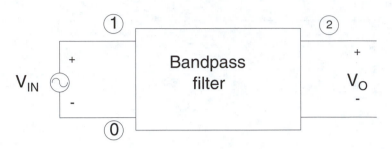

FIGURE 2.17
Block diagram of a filter circuit.

Example 2.9   Laplace Transform Description of a Bandpass Filter

The voltage transfer function of a second-order bandpass filter (Figure 2.17) is

$$\frac{V_{out}}{V_{in}} = \frac{As}{s^2 + Bs + C} \tag{2.11}$$

where

s is the Laplace transform variable and A, B, and C are expressions describing the filter characteristics.

If $A = R/L$, $B = R/L$, $C = 1/LC$, L = 5 H, R = 100 Ω, and C = 10 μF, plot the magnitude response.

Solution

PSPICE program:

```
FREQUENCY RESPONSE OF A FILTER
VIN 1  0   AC 1
.AC DEC 20 1 10K;
*FILTER CONSTANTS
.PARAM A = {100/5}
.PARAM B = {100/5}
.PARAM C = {1/5.0E-6}
*FILTER TRANSFER FUNCTION
EBANPAS 2 0 LAPLACE {V(1,0)} = {A*S/(S*S+S*B+C)}
.PROBE V(2) V(1); FILTER OUTPUT
.END
```

The magnitude response is shown in Figure 2.18.

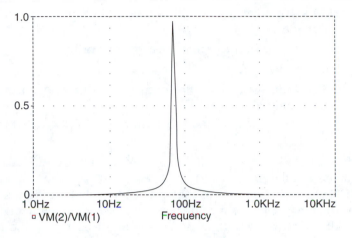

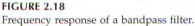

FIGURE 2.18
Frequency response of a bandpass filter.

2.7   Monte Carlo Analysis (.MC)

The parameters of electric element and electronic devices vary due to tolerances incurred from manufacturing processes and also due to aging of components. Monte Carlo analysis allows the user to vary device parameters
and to observe the overall system for variations in circuit parameters. The
general form of the Monte Carlo analysis statement is

    **.MC NUM_RUNS ANALYSIS OUTPUT_VARIABLE FUNCTION
OPTIONS [SEED = VALUE]**

where
    .MC statement causes Monte Carlo (statistical) analysis of a circuit to be
        done.
    NUM_RUNS is the number of runs of the selected analysis (DC, AC, or
        transient). The first run is done with nominal values of all components.
        Subsequent runs are done with variations of **DEV** and **LOT** tolerances
        on each **.MODEL** parameter. For printed results, upper limit of
        NUM_RUNS is 2000. However, for PROBE results, the limit is 400.
    ANALYSIS is the analysis type that must be one of these: DC, AC, or
        TRAN. The specified analysis type is repeated in the *subsequent* passes
        of the analysis.
    OUTPUT_VARIABLE is the output variable that is to be tested. It is iden-
        tical in format to that of a .PRINT output variable, discussed in
        Section 1.7.
    FUNCTION specifies the operation to be performed on the
        OUTPUT_VARIABLE in order to reduce the values of the latter to a sin-
        gle value. The function must be one of the following.

1. YMAX: the greatest difference in each waveform from the nominal value

2. MAX: the maximum value of each waveform

3. MIN: the minimum value of each waveform

4. RISE_EDGE<value>: the first occurrence of the waveform above a threshold value

5. FALL_EDGE <value>: the first occurrence of the waveform below a threshold value

It should be noted that FUNCTION has no effect on the PROBE data saved from the simulation.

OPTIONS are additional items that can be requested during the Monte Carlo analysis. Options include none or more of the following:

1. LIST: will print out, at the beginning of each run, the model parameter values used during each run

2. OUTPUT (output_type): request output from runs subsequent to the nominal (first) run. Output_type can be one of the following:

   a. ALL: all outputs for all runs are generated, including the nominal runs

   b. FIRST<value>: generates outputs only for the first runs where *n* is specified by <value>

   c. EVERY<value>: generates output every *n*th value, where *n* is specified by <value>

   d. RUNS<value>: does analysis and generates output only for specified runs; given by <value> up to 25 values may be specified in a list

   e. RANGE(<low_value>,<high_value>): restricts the range over which <FUNCTION> will be evaluated. An "*" can be used in place of either <low_value> or <high_value> to indicate for all values If RANGE is omitted, then <FUNCTION> is evaluated over the whole sweep range. This is equivalent to RANGE(*,*).

[SEED = VALUE]: the seed value for the random number generator within the Monte Carlo analysis. The default value is 17,533. For almost all analyses, it is advisable to use the default seed value to achieve a constant set of results.

2.7.1 Component Tolerances for Monte Carlo Analysis

In Monte Carlo analysis, device parameters are allowed to change. There are two ways of changing device parameters in PSPICE: (1) deviation of device parameters **(DEV)** and (2) deviation of lot parameters **(LOT)**. The **DEV** parameter allows components to vary independently of other components. The **LOT** parameter allows components from the same lot to track each other.

The **.MODEL** statement of PSPICE is used to assign the tolerances of components. The **.MODEL** statement was discussed in Section 2.1. The general format of including the **DEV** and **LOT** parameters in the **.MODEL** statement is

[DEV/DISTRIBUTION] <VALUE> [%]
[LOT/DISTRIBUTION] <VALUE> [%]

where
   DISTRIBUTION parameters can be one of the following
   1. UNIFORM: generates uniform distributed deviations over the range ± <value>
   2. GAUSS: generates deviations with a Gaussian distribution over the range ±3σ and <value> specifies the ±1σ deviation. PSPICE limits the values of Gaussian distributions to ±4σ, where σ is the standard deviation.
   3. USER_DEFINED DISTRIBUTION: generates deviations using a user_defined distribution and <value> specifies the ±1 deviation in the user_defined_distribution.

   VALUE parameter is the deviation of the component value in percentages.

The following examples illustrate how device parameters can be changed in PSPICE.

1. *DEV tolerance used with .MODEL statement:*

   R1 1 2 RMOD1 10K
   R2 2 3 RMOD1 50K
   .MODEL RMOD1 RES(R=1 DEV=10%)

   During statistical analysis runs, the values of R1 and R2 are varied at most by 10%. The variations in R1 and R2 will be independent. In the simulation, R1 can take any value from 9K and 11K, and R2 any values from 45K to 55K.

2. *LOT tolerance used with .MODEL statement:*

   R3 3 4 RMOD1 15K
   R4 4 5 RMOD! 20K
   .MODEL RMOD1 RES(R=1 LOT=5%)

   During statistical analysis run, the values of R3 and R4 are varied by at most 5%. However, R3 and R4 will increase or decrease by the same percentage.

3. *DEV and LOT tolerances used with .MODEL statement:*

C1 10 11 CMOD 50nF
C2 11 12 CMOD 100nF
.MODEL CMOD CAP(C=1 LOT=1% DEV=5%)

During simulations, C1 and C2 are assigned LOT variations up to
1% and DEV variations up to 5%. The two tolerances add. Thus,
C1 and C2 can be up to 6% from their nominal values. The 5%
tolerance in DEV makes the changes in C1 and C2 uncorrelated,
but the 1% tolerance in LOT C1 and C2 makes the change in C1
and C2 correlated.

2.7.2   Simulation

There is a trade-off between the quantities of statistical data obtained from
the Monte Carlo analysis versus the simulation time. To obtain a realistic
estimation of the true maximum and minimum limits of component or
system variation, it is necessary to perform several runs of the Monte Carlo
analysis. However, it takes longer to do more runs. The simulation time is
proportional to the number of runs to be performed. The user has to make
intelligent decisions as to the number of runs and the time needed for a
simulation.

Three types of data are available from a Monte Carlo analysis.

1. The .OUT file contains model parameters with tolerances applied.
2. Using PROBE, .PRINT, and .PLOT, the waveforms for each run are
   available for viewing, printing, or plotting.
3. The .OUT file may contain the summary of all the runs and may
   be obtained using the Monte Carlo FUNCTION statement
   described in Section 2.7.

The following example shows the application of the Monte Carlo analysis.

Example 2.10   Monte Carlo Analysis of a Bipolar Transistor Biasing Network

A universal bipolar biasing network is shown in Figure 2.19. RB = 10 K, RE =
1 K, RC = 1 K, VCC = 10 V, and VEE = −10 V. The resistors have 5% tolerance
with uniform distribution. If the beta of the transistor, β_f, is 100 and the
device variation is 10% with uniform distribution, find the changes in the
biasing point.

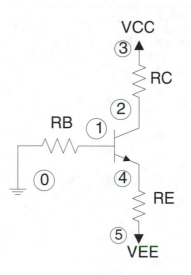

FIGURE 2.19
Universal biasing network.

Solution

PSPICE program:

```
MONTE CARLO ANALYSIS
*CIRCUIT ELEMENTS
VCC 3   0   DC 10V
VEE 5   0   DC -10V
RB  1   0   RMOD   10K
RC  3   2   RMOD   1K
RE  4   5   RMOD   1K
Q1  2   1   4   QMOD
*MODEL OF DEVICE WITH TOLERANCES
.MODEL RMOD RES(R=1 DEV/UNIFORM 5%)
.MODEL QMOD NPN (BF=100 DEV/UNIFORM 10% VJC=0.7V)
*MONTE CARLO ANALYSIS
.MC 100 DC I(RC) MAX LIST OUTPUT ALL
*100 RUNS, MONITOR CURRENT THROUGH RC USING YMAX
COLLATING FUNCTION
.DC VCC 10 10 1
.PRINT DC V(2, 4) I(RC)
.END
```

The edited version of the PSPICE output file is shown in Table 2.13.

TABLE 2.13

Edited Results of Monte Carlo Analysis

```
MONTE CARLO NOMINAL

**** CURRENT MODEL PARAMETERS FOR DEVICES REFERENCING RMOD
                      RB              RC              RE
            R    1.0000E+00     1.0000E+00   1.0000E+00

**** CURRENT MODEL PARAMETERS FOR DEVICES REFERENCING QMOD
                      Q1
            BF    1.0000E+02

MONTE CARLO ANALYSIS
VCC          V(2,4)       I(RC)
1.000E+01   3.393E+00   8.262E-03
*************************************************************

MONTE CARLO ANALYSIS MONTE CARLO PASS 2

**** CURRENT MODEL PARAMETERS FOR DEVICES REFERENCING RMOD
                      RB              RC              RE
            R    9.5672E-01     9.6940E-01   1.0304E+00

**** CURRENT MODEL PARAMETERS FOR DEVICES REFERENCING QMOD
                      Q1
            BF    9.0106E+01

**** DC TRANSFER CURVES   TEMPERATURE = 27.000 DEG C
VCC          V(2,4)       I(RC)
1.000E+01   3.931E+00   7.989E-03
*************************************************************

MONTE CARLO ANALYSIS MONTE CARLO PASS 100

**** UPDATED MODEL PARAMETERS   TEMPERATURE = 27.000 DEG C

**** CURRENT MODEL PARAMETERS FOR DEVICES REFERENCING RMOD
                      RB              RC              RE
            R    1.0143E+00     1.0146E+00   1.0410E+00
```

```
**** CURRENT MODEL PARAMETERS FOR DEVICES REFERENCING QMOD
                            Q1
              BF     9.0247E+01


**** DC TRANSFER CURVES    TEMPERATURE   =   27.000 DEG C
VCC            V(2,4)       I(RC)
1.000E+01  3.724E+00    7.874E-03
************************************************************

MONTE CARLO ANALYSIS

**** SORTED DEVIATIONS OF I(RC)    TEMPERATURE  =  27.000 DEG C

                     MONTE CARLO SUMMARY
************************************************************
RUN MAXIMUM VALUE

Pass 18    8.6788E-03 at VCC = 10
           (105.04% of Nominal)

Pass 31    8.6637E-03 at VCC = 10
           (104.86% of Nominal)

NOMINAL    8.2623E-03 at VCC = 10

Pass 99    8.2504E-03 at VCC = 10
           (99.855% of Nominal)

Pass 61    7.8950E-03 at VCC = 10
           (95.554% of Nominal)

Pass 100   7.8736E-03 at VCC = 10
           (95.295% of Nominal)
```

When the analysis is performed, the first run is done using the nominal values of the devices. The results give the values for RB, RC, RE, and transistor beta for several runs. For each run, the values for VCC, node voltage V(2, 4), and current I(RC) are given. In addition, the summary of the Monte Carlo analysis shows the maximum value and its percentage change of the nominal value for the various runs.

2.8   Sensitivity and Worst-Case Analysis (.WCASE)

Critical elements in a circuit can be determined using the **.WCASE** statement. During the WCASE analysis, only one element is varied per run. This allows PSPICE to calculate the sensitivity of an output variable for each element parameter in the circuit. Once the sensitivity of each element is known, one final run is done with all the component parameters; this will give the worst-case output.

The component parameters that are varied in the sensitivity and worst-case analysis are specified by the DEV and LOT tolerances of each .MODEL parameter. This was discussed in Section 2.7. The general format for the **.WCASE** statement is

.WCASE ANALYSIS OUTPUT_VARIABLE FUNCTION [OPTIONS]

where

.WCASE causes the sensitivity and worst-case analysis to be performed.

ANALYSIS is one of the following analysis types: DC, AC, or TRAN. All the analysis types specified in the circuit file are performed during the first nominal run. Only the selected analysis, specified in the **.WCASE** statements, is re-performed during the subsequent runs.

OUTPUT_VARIABLE is the output variable to be monitored. Its format is identical to that of the .PRINT output variable discussed in Section 1.7.

FUNCTION specifies the operation to be performed on the output variable to reduce the values of the output variables to a single value. The function must be one of the following:

YMAX: finds the absolute value of the *greatest difference* in each waveform from the nominal run

MAX: finds the maximum value of each waveform

MIN: finds the minimum value of each waveform

RISE_EDGE(<VALUE>): finds the *first occurrence* of the waveform crossing *above* the threshold <value>. The waveform must have one or more points at or below <value>, followed by one above. The output value listed will be where the waveform increases above <value>

FALL_EDGE <value>: finds the first occurrence of the waveform crossing below the threshold <value>. The waveform must have one or more points at or above <value>, followed by one below. The output value listed will be where the waveform decreases below <value>

OPTION includes none or more of the following:

LIST: will print the updated model parameters for sensitivity analysis.

OUTPUT ALL: requests output from the sensitivity runs, after the nominal (first) run. The output from any run is governed by the .PRINT, .PLOT, and .PROBE statements in the file. If OUTPUT ALL is omitted, then only the nominal and worst-case runs produce output. OUTPUT ALL will ensure that all sensitivity information is saved for PROBE.

RANGE (<low value>, <high value>): restricts the range over which FUNCTION will be evaluated. An "X" can be used in place of <value> to indicate for all values. If RANGE is omitted, then FUNCTION is evaluated over the whole sweep range. This is equivalent to RANGE(*,*).

HI or LOW: specifies which direction the worst-case run is to go (relative to the nominal). If FUNCTION is YMAX or MAX, the default is HI. Otherwise, the default is LOW.

VARY DEV/VARY LOT/VARY BOTH: by default, any device that has a model parameter specifying either a DEV tolerance or a LOT tolerance will be included in the analysis. You may limit the analysis to only those devices that have DEV or LOT tolerance by specifying the appropriate option. The default is VARY BOTH. When VARY BOTH is used, sensitivity to parameters with both DEV and LOT specifications is checked only with respect to LOT variations. The parameter is then maximized utilizing both DEV and LOT tolerances. All devices referencing the model will have the same parameter value for the worst-case simulation.

DEVICES (list of device types): by default, all devices are included in the sensitivity and worst-case analysis. One can limit the devices considered by listing the device types after the keyword DEVICES.

It should be noted that <function> and all [options] do not affect PROBE data obtained from the simulation. They are applicable to the output file. The following example will explore the worst-case analysis statement for two amplifier circuits.

Example 2.11 Worst-Case and Sensitivity Analysis of an Instrumentation Amplifier

For the instrumentation amplifier shown in Figure 2.20, R1 = 1 KΩ, R2 = R3 = 10 KΩ, R4 = R5 = 20 KΩ, and R6 = R7 = 100 KΩ. If the resistors have tolerance of 5%, find the sensitivity and the worst-case gain of the amplifier. The source voltage VIN has an amplitude of 1 mV and a frequency of 5 KHz. Assume the op amp has an input resistance of 10^{12} Ω, an open loop gain of 10^7, and zero output resistance.

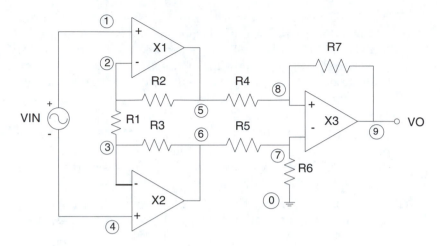

FIGURE 2.20
Instrumentation amplifier.

Solution

PSPICE program:

```
INSTRUMENTATION AMPLIFIER
.OPTIONS RELTOL=0.05;5% COMPONENTS (SENSITIVITY RUN)
*
VIN 1   4   AC 1E-3; INPUT SIGNAL
.AC LIN 10 1 5KHZ; FREQUENCY OF SOURCE AND AC ANALYSIS
X1   2   1   5   OPAMP; OP AMP X1
X2   3   4   6   OPAMP; OP AMP X2
X3   8   7   9   OPAMP; OP AMP X3
*RESISTORS WITH MODELS
R1   2   3   RMOD1 1K
R2   2   5   RMOD1 10K
R3   3   6   RMOD1 10K
R4   5   8   RMOD1 20K
R5   6   7   RMOD1 20K
R6   7   0   RMOD1 100K
R7   8   9   RMOD1 100K
.MODEL RMOD1 RES(R=1 DEV=5%); 5% RESISTORS
*
.WCASE AC V(9) MAX OUTPUT ALL; SENSITIVITY & WORST CASE
```

(continued)

```
*
.PROBE V(9)
*SUBCIRCUIT
.SUBCKT OPAMP 1 2 3
* - INPUT; + INPUT; OUTPUT
RIN 1 2 1.0E12
EVO 0 3 1 2 1.0E7
.ENDS OPAMP
.END
```

Table 2.14 shows the output file for the sensitivity analysis and Table 2.15 shows the worst-case results from the simulation.

TABLE 2.14

Sensitivity Analysis of an Instrumentation Amplifier

```
INSTRUMENTATION AMPLIFIER

**** SORTED DEVIATIONS OF V(9)     TEMPERATURE = 27.000 DEG C

                       SENSITIVITY SUMMARY
* * * * * * * * * * * * * * * * * * * * * * * * * * * * * * * * * * * * * * * * * * * * * * * * * * * * * * * *
RUN          MAXIMUM VALUE

R7 RMOD1 R  .1098 at F =   5.0000E+03
             (.9167% change per 1% change in Model Parameter)

R2 RMOD1 R  .1075 at F =   5.0000E+03
             (.4762% change per 1% change in Model Parameter)

R3 RMOD1 R  .1075 at F =   5.0000E+03
             (.4762% change per 1% change in Model Parameter)

R6 RMOD1 R  .1054 at F =   5.0000E+03
             (.08 % change per 1% change in Model Parameter)

NOMINAL     .105 at F =   5.0000E+03

R5 RMOD1 R  .1046 at F =   5.0000E+03
             (-.0826% change per 1% change in Model Parameter)

R4 RMOD1 R  .1004 at F =   5.0000E+03
             (-.873 % change per 1% change in Model Parameter)

R1 RMOD1 R  .1002 at F =   5.0000E+03
             (-.907 % change per 1% change in Model Parameter)
```

TABLE 2.15

Worst Case Results

```
INSTRUMENTATION AMPLIFIER

**** WORST CASE ANALYSIS      TEMPERATURE = 27.000 DEG C
                WORST CASE ALL DEVICES
********************************************************

**** UPDATED MODEL PARAMETERS  TEMPERATURE = 27.000 DEG C
                WORST CASE ALL DEVICES
********************************************************

DEVICE          MODEL       PARAMETER       NEW VALUE

R1              RMOD1        R               .95
(Decreased)

R2              RMOD1        R               1.05
(Increased)

R3              RMOD1        R               1.05
(Increased)

R4              RMOD1        R               .95
(Decreased)

R5              RMOD1        R               .95
(Decreased)

R6              RMOD1        R               1.05
(Increased)

R7              RMOD1        R               1.05
(Increased)
********************************************************

*** SORTED DEVIATIONS OF V(9)   TEMPERATURE = 27.000 DEG C
                WORST CASE SUMMARY
********************************************************

RUN             MAXIMUM VALUE

ALL DEVICES     .1277 at F =   5.0000E+03
                (121.61% of Nominal)

NOMINAL         .105 at F =    5.0000E+03
********************************************************
```

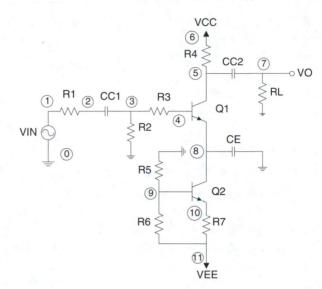

FIGURE 2.21
Current-biased common
emitter amplifier.

In Table 2.14, the nominal value of the voltage V(9) at a frequency of 5 KHz is 0.105 V. Results for the output voltage are given for changes in R1 to R7. The output voltage is sorted with the maximum value of V(9) of 0.1098 V listed first (caused by the change in R7) and the minimum value of V(9) of 0.1002 V listed last (caused by the change in R1).

Table 2.15 shows that R1, R4, and R5 are set to their minimum allowed values and that R2, R3, R6, and R7 are set their maximum allowed values. These values condition the output voltage V(9) to be maximum. The maximum voltage at node 9, considering all devices, is 0.1277 V. This maximum value is 121.61% of the nominal value of 0.105 V.

Example 2.12 Worst-Case and Sensitivity Analysis of a Current-Biased Common Emitter Amplifier

For the common emitter amplifier shown in Figure 2.21, R1 = 50 Ω, R2 = 1 KΩ, R3 = 10 KΩ, R4 = 1 KΩ, R5 = 16 KΩ, R6 = 20 KΩ, R7 = RL = 10 KΩ, VCC = 5 V, VEE = –5 V, CC1 = CC2 = 20 μF, and CE = 100 μF. Q1 and Q2 are transistors Q2N2222. The tolerance of the resistors is 5% and that of capacitors is 5%. The transistor has gain BF of 100 with device variation of 30% (uniform distribution). What is the worst-case output voltage?

Solution

PSPICE program:

```
COMMON-EMITTER AMPLIFIER
.OPTIONSRELTOL = 0.05;
VIN 1  0  AC 1E-3;AC INPUT SIGNAL
```

(continued)

```
*RESISTORS WITH MODEL
R1   1   2    RMOD2   50
R2   3   0    RMOD2   1K
R3   3   4    RMOD2   10K
R4   6   5    RMOD2   1K
R5   9   0    RMOD2   16K
R6   9   11   RMOD2   20K
R7   10  11   RMOD2   10K
RL   7   0    RMOD2   10K
.MODEL RMOD2 RES(R=1 DEV=5%); 5% RESISTORS
*
*CAPACITORS WITH MODEL
CC1  3   2    CMOD2   20E-6
CE   8   0    CMOD2   100E-6
CC2  5   7    CMOD2   20E-6
.MODEL CMOD2 CAP(C=1 DEV=5%); 5% CAPACITORS
*SOURCE VOLTAGES
VCC  6   0    DC       10V
VEE  11  0    DC      -10V
* TRANSISTOR WITH MODELS
Q1   5   4   8        Q2N2222
Q2   8   9   10       Q2N2222
.MODEL Q2N2222 NPN (BF = 100 DEV/UNIFORM 30% IS=3.295E-14
VA=200)
.AC LIN 1 1KHZ 1KHZ;   FREQUENCY OF SOURCE & AC ANALYSIS
* SENSITIVITY AND WORSE CASE ANALYSIS FOR AC ANALYSIS
.WCASE AC V(7)) MAX OUTPUT ALL; FOR AC ANALYSIS
.END
```

Table 2.16 shows edited sensitivity summary report. In addition, Table 2.17 shows the worst-case summary of results.

TABLE 2.16

Edited Sensitivity Summary Report for Example 2.12

```
*********************************************************
COMMON-EMITTER AMPLIFIER
**** SORTED DEVIATIONS OF V(7)   TEMPERATURE = 27.000 DEG C
                    SENSITIVITY SUMMARY
*********************************************************
```

(continued)

```
RUN MAXIMUM VALUE

R4 RMOD2 R   6.0339E-03 at F =   1.0000E+03
               (.8945% change per 1% change in Model
               Parameter)

Q1 Q2N2222 BF5.9564E-03 at F =   1.0000E+03
               (.626 % change per 1% change in Model
               Parameter)

R6 RMOD2 R   5.8245E-03 at F =   1.0000E+03
               (.1693% change per 1% change in Model
               Parameter)

RL RMOD2 R   5.8012E-03 at F =   1.0000E+03
               (.0885% change per 1% change in Model
               Parameter)

R2 RMOD2 R   5.7868E-03 at F =   1.0000E+03
               (.0387% change per 1% change in Model
               Parameter)

NOMINAL       5.7756E-03 at F =   1.0000E+03

Q2 Q2N2222 BF5.7755E-03 at F =   1.0000E+03
               (-490.2E-06% change per 1% change in Model
               Parameter)

CE CMOD2 C   5.7737E-03 at F =   1.0000E+03
               (-6.5081E-03% change per 1% change in Model
               Parameter)

CC1 CMOD2 C  5.7737E-03 at F =   1.0000E+03
               (-6.5516E-03% change per 1% change in Model
               Parameter)

CC2 CMOD2 C  5.7737E-03 at F =   1.0000E+03
               (-6.6177E-03% change per 1% change in Model
               Parameter)

R1 RMOD2 R   5.7592E-03 at F =   1.0000E+03
               (-.057 % change per 1% change in Model
               Parameter)

R5 RMOD2 R   5.7233E-03 at F =   1.0000E+03
               (-.1813% change per 1% change in Model
               Parameter)
```

```
R7 RMOD2 R   5.6742E-03 at F =   1.0000E+03
             (-.3513% change per 1% change in Model
             Parameter)

R3 RMOD2 R   5.5953E-03 at F =   1.0000E+03
             (-.6244% change per 1% change in Model
             Parameter)
*********************************************************
```

TABLE 2.17

Edited Worst-Case Summary for Example 2.12

```
*********************************************************
COMMON-EMITTER AMPLIFIER

**** UPDATED MODEL PARAMETERS  TEMPERATURE = 27.000 DEG C

                WORST CASE ALL DEVICES
*********************************************************
```

DEVICE	MODEL	PARAMETER	NEW VALUE
CC1 (Decreased)	CMOD2	C	.95
CE (Decreased)	CMOD2	C	.95
CC2 (Decreased)	CMOD2	C	.95
Q1 (Increased)	Q2N2222	BF	130
Q2 (Decreased)	Q2N2222	BF	70
R1 (Decreased)	RMOD2	R	.95
R2 (Increased)	RMOD2	R	1.05
R3 (Decreased)	RMOD2	R	.95

R4	RMOD2	R	1.05
(Increased)			
R5	RMOD2	R	.95
(Decreased)			
R6	RMOD2	R	1.05
(Increased)			
R7	RMOD2	R	.95
(Decreased)			
RL	RMOD2	R	1.05
(Increased)			

```
* * * * * * * * * * * * * * * * * * * * * * * * * * * * * * * * * * * * * * * * * * * * * * * * * * *

                    WORST CASE ALL DEVICES

COMMON-EMITTER AMPLIFIER

**** SORTED DEVIATIONS OF V(7)   TEMPERATURE = 27.000 DEG C

                    WORST CASE SUMMARY
* * * * * * * * * * * * * * * * * * * * * * * * * * * * * * * * * * * * * * * * * * * * * * * * * * *

RUN                 MAXIMUM VALUE

ALL DEVICES         7.6507E-03 at F =   1.0000E+03
                    (132.47% of Nominal)

NOMINAL             5.7756E-03 at F =   1.0000E+03
* * * * * * * * * * * * * * * * * * * * * * * * * * * * * * * * * * * * * * * * * * * * * * * * * * *
```

In Table 2.16, the nominal value of the voltage V(7) at a frequency of 1 KHz is 5.7756 mV. Results for the output voltage are given for changes in R1 to RL, CC1, CC2, CE, and also the changes in BF of transistors Q1 and Q2. The output voltage is sorted with the maximum value of V(7) being 6.0339 mV listed first (caused by the change in R4) and the minimum value of V(7) being 5.5953 mV listed last (caused by the change in R3).

Table 2.17 shows that CC1, CE, CC2, R1, R3, R5, R7, and BF of Q2 are set to their minimum allowed values, and R2, R4, R6, RL and BF of Q1 are set their maximum allowed values. These values condition the output voltage V(7) to be maximum. The maximum voltage at node 7, considering all devices, is 7.6507 mV. This maximum value is 132.47% of the nominal value of 5.7756 mV.

2.9   Fourier Series (.FOUR)

The periodic signal $g(t)$ can be expressed as an infinite series of sine and cosine terms; that is,

$$v(t) = \frac{a_0}{2} + \sum_{n=1}^{\infty} a_n \cos(nw_0 t) + b_n \sin(nw_0 t) \tag{2.12}$$

where

$$w_0 = \frac{2\pi}{T_P} \tag{2.13}$$

$$T_P = \text{period of } v(t)$$

The Fourier coefficients a_n and b_n are determined by the following equations:

$$a_n = \frac{2}{T_P} \int_{t_0}^{t_0 + T_P} v(t) \cos(n\omega_0 t)dt \qquad n = 0, 1, 2, 3, \dots \tag{2.14}$$

$$b_n = \frac{2}{T_P} \int_{t_0}^{t_0 + T_P} v(t) \sin(n\omega_0 t)dt \qquad n = 0, 1, 2, 3, \dots \tag{2.15}$$

Equation (2.12) can be rewritten as

$$v(t) = C_0 + \sum_{n=1}^{\infty} c_n \sin(n\omega_0 t + \theta_n)dt \tag{2.16}$$

where

$$C_0 = \frac{a_0}{2} \tag{2.17}$$

$$C_n = \sqrt{a_n^2 + b_n^2} \tag{2.18}$$

and

$$\theta_n = \tan\left(\frac{a_n}{b_n}\right) \tag{2.19}$$

C_0 is the dc component and C_n is the coefficient for the nth harmonic. The first harmonic is obtained with $n = 1$. The latter is also called the fundamental with fundamental frequency of w_0. When $n = 2$, we have the second harmonic,

$n = 3$ the third harmonic, and so on. Fourier series representation in PSPICE is shown in Equation (2.16).

PSPICE allows the user to do Fourier series analysis using the .FOUR statement. PSPICE calculates the signal Fourier components from dc to the nth component. The amplitude and phase of the harmonic components are tabulated in the circuit output file. The .FOUR statement results are automatically printed and it does not require .PRINT, .PLOT, or .PROBE statements. The general form of the .FOUR statements is

.FOUR FUNDA_FREQUENCY NUMBER_OF_HARMONICS OUTPUT_VARIABLE

where

FUNDA_FREQUENCY is the fundamental frequency of the periodic waveform in hertz.

NUMBER_OF_HARMONICS is the number of harmonics to be calculated. The dc component, fundamental, 2nd to 9th harmonics are calculated by default. More harmonics can be obtained by specifying the number of harmonics

OUTPUT_VARIABLE is an output variable of the same form as in a .PRINT statement or .PLOT statement for transient analysis.

The .FOUR statement requires a .TRAN statement. The Fourier series coefficients are obtained by PSPICE by performing a Fourier integral on selected output of transient analysis results at evenly spaced time points. The interval used is *print step size* in the .TRAN statement, or 1% of the *final time value* (TSTOP) if smaller. A second-order polynomial interpolation is employed to calculate the transient analysis output values used in the integration. Not all the transient analysis results are used by the .FOUR statement. Only the interval from the end, back to $1/$(*fundamental frequency*) before the end is used. This means that the transient analysis must be at least $1/$(fundamental frequency) seconds long. The following example illustrates the use of the .FOUR statement.

Example 2.13   Fourier Series Expansion of a Half-Wave Rectifier

Figure 2.22 shows a circuit for half-wave rectification. For a sinusoidal input with a frequency of 500 Hz, amplitude of 1 V, find the frequency content at the output voltage VO.

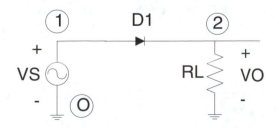

FIGURE 2.22
Half-wave rectifier.

Solution

PSPICE program:

```
*HALF WAVE RECTIFIER
VS  1  0   SIN(0 1 500 0 0)
D1  1  2   DMOD
.MODEL DMOD D
RL  2  0   1K
*CONTROL STATEMENTS
.TRAN 1E-6 4E-3
.FOUR 500 V(2)
.END
```

The output of the program is shown in Table 2.18.

TABLE 2.18

Fourier Components of an Output of a Half-Wave Rectifier

```
*HALF WAVE RECTIFIER

****    FOURIER ANALYSIS                    TEMPERATURE = 27.000 DEG C
**********************************************************************

FOURIER COMPONENTS OF TRANSIENT RESPONSE V(2)

DC COMPONENT = 7.341440E-02
```

HARMONIC NO	FREQUENCY (HZ)	FOURIER COMPONENT	NORMALIZED COMPONENT	PHASE (DEG)	NORMALIZED PHASE (DEG)
1	5.000E+02	1.333E-01	1.000E+00	-1.094E-02	0.000E+00
2	1.000E+03	9.815E-02	7.365E-01	-9.002E+01	-9.001E+01
3	1.500E+03	5.517E-02	4.140E-01	1.800E+02	1.800E+02
4	2.000E+03	1.881E-02	1.412E-01	9.002E+01	9.003E+01
5	2.500E+03	2.417E-03	1.814E-02	1.787E+02	1.787E+02
6	3.000E+03	8.469E-03	6.355E-02	8.953E+01	8.954E+01
7	3.500E+03	5.333E-03	4.002E-02	-2.022E-01	-1.913E-01
8	4.000E+03	2.029E-04	1.523E-03	-7.174E+01	-7.173E+01
9	4.500E+03	2.472E-03	1.855E-02	-2.385E+00	-2.374E+00

```
TOTAL HARMONIC DISTORTION = 8.603263E+01 PERCENT
```

From the results of Table 2.18, the output voltage can be written as:

$$vo(t) = 0.073 + 0.133\sin(1000\pi t - 0.011°) + 0.098\sin(2000\pi t - 90.02°)$$
$$+ 0.055\sin(3000\pi t + 180°) + 0.019\sin(4000\pi t + 90.02°) + \ldots$$

(2.20)

2.9.1 Fourier Analysis Using PROBE

In the previous section, the Fourier series expansion was obtained in tabular form using both .TRAN and .FOUR statements in the input file. The .OUT file contains the tabulated Fourier components. One can also use the PROBE to obtain the Fourier components of a signal in a graphic format. This is achieved by:

- Having .TRAN and .PROBE statements as part of input file
- Plotting, using probe to view the transient analysis results of the required waveform
- Selecting "X-axis" from probe menu
- Choosing "Fourier" from sub-menu

To reduce excessive data points saved to memory, it is important to specify the required output variables in the .PROBE statement. PROBE uses the Fast Fourier Transform (FFT) algorithm to obtain the Fourier components. FFT requires that (1) the time intervals between the individual samples be equally spaced and (2) the number of data points should be a power of two. Before the Fourier components are calculated, PROBE creates a new set of data points (power of 2) based on the points created by PSPICE using the interpolation method.

To illustrate the Fourier analysis using PROBE, we redo Example 2.13. The netlist is shown below.

```
*HALF WAVE RECTIFIER
VS   1   0   SIN(0 1 500 0 0)
D1   1   2   DMOD
.MODEL DMOD D
RL   2   0   1K
*CONTROL STATEMENTS
.TRAN 1E-6 4E-3
.FOUR 500 V(2)
.PROBE V(2)
.END
```

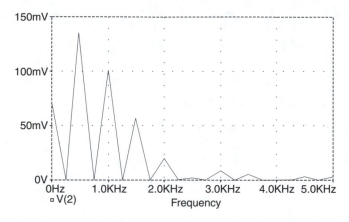

FIGURE 2.23
Frequency plot of half-wave rectifier.

The Fourier coefficients at output of the rectifier is displayed using PROBE. Select "X-axis" from the PROBE menu and then select "Fourier" from the sub-menu. The output obtained is shown in Figure 2.23.

2.9.2 RMS and Harmonic Distortion

Because the Fourier series expansion decomposes a periodic signal into the sum of sinusoids, the root-mean-squared (rms) value of the periodic signal can be obtained by adding the rms value of each harmonic component vectorially; that is,

$$V_{rms} = \sqrt{V_{1,\,rms}^2 + V_{2,\,rms}^2 + V_{3,\,rms}^2 + \ldots + V_{n,\,rms}^2} \qquad (2.21)$$

where
    V_{rms} = rms value of the periodic signal
    $V_{1,rms}, V_{2,rms}, \ldots, v_{n,rms}$ are rms values of the harmonic components

From Equations (2.16) and (2.21), the rms value of $v(t)$ is

$$V_{rms} = \sqrt{C_0^2 + \left(\frac{C_1}{\sqrt{2}}\right)^2 + \left(\frac{C_2}{\sqrt{2}}\right)^2 + \ldots + \left(\frac{C_n}{\sqrt{2}}\right)^2} \qquad (2.22)$$

Harmonic distortion is a measure of the distortion a signal undergoes as it passes through a network (such as an amplifier, filter, or transmission line). Harmonic distortion can also show the discrepancy between an approximation of a signal (obtained from synthesizing sinusoidal components) and its actual waveform. The smaller the harmonic distortion, the more nearly the approximation of a signal resembles the true signal. For the Fourier series expansion, the percent distortion for each individual component is given as:

$$\text{Percent distortion for } n\text{-th harmonic} = \frac{C_n}{C_1} * 100 \qquad (2.23)$$

where

C_n is the amplitude of the nth harmonic

C_1 is the amplitude of the fundamental harmonic

The total harmonic distortion (THD) involves all the frequency components and it is given as

$$\text{Percent THD} = \sqrt{\left(\frac{C_2}{C_1}\right)^2 + \left(\frac{C_3}{C_1}\right)^2 + \cdots + \left(\frac{C_n}{C_1}\right)^2} * 100\% \qquad (2.24)$$

or

$$\text{Percent THD} = \frac{\sqrt{C_2^2 + C_3^2 + \cdots + C_n^2}}{C_1} * 100\% \qquad (2.25)$$

The following example determines the total harmonic distortion at the output of an RC network.

Example 2.14   Square Wave Signal through Two-Stage RC Network

Figure 2.24(a) shows a two-stage RC network. C1 = C2 = 1 μF and R1 = R2 = 1 KΩ. The periodic square wave, shown in Figure 2.24(b), is applied to the input of the RC network. Find the frequency content at the output. Determine the rms value of the output signal and calculate the total harmonic distortion of the output signal.

Solution

PSPICE program:

```
*SQUARE WAVE SIGNAL THROUGH NETWORK
VS          1       0       PULSE(-10 10 0 1E-6 1E-6 1E-3 2E-3)
C1          1       2       1E-6
R1          2       0       1E3
C2          2       3       1E-6
R2          3       0       1E3
.TRAN       5E-6    4E-3
.FOUR       500 V(1) V(3)
.PROBE      V(1)    V(3)
.END
```

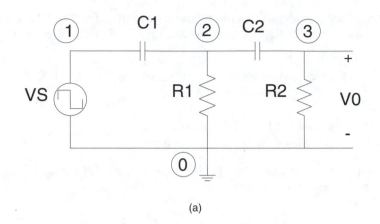

(a)

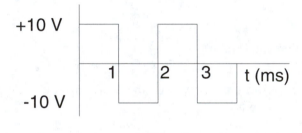

(b)

FIGURE 2.24
(a) A two-stage RC network and (b) input signal to RC network.

The Fourier coefficients for the first nine harmonics for output waveform voltage are shown in Table 2.19.

The rms value of the output voltage is obtained using Equation (2.22) and it is given as

$$V_{rms} = \sqrt{(0.461)^2 + \left(\frac{9.608}{\sqrt{2}}\right)^2 + \left(\frac{0.0622}{\sqrt{2}}\right)^2 + \left(\frac{4.046}{\sqrt{2}}\right)^2 + \dots \left(\frac{1.39}{\sqrt{2}}\right)^2}$$

$$= 7.7588 \text{ V}$$

The percent total harmonic distortion is given by Equation (2.24) and its value is

$$\text{Percent THD} = \frac{\sqrt{(0.0622)^2 + (4.046)^2 + (0.0314)^2 + (2.484)^2 + \dots + (1.390)^2}}{9.608}$$

$$= 54.74\%$$

TABLE 2.19

Fourier Components for Output Voltage of Figure 2.24

```
*SQUARE WAVE SIGNAL THROUGH NETWORK

**** FOURIER ANALYSIS                      TEMPERATURE = 27.000 DEG C
****************************************************************

FOURIER COMPONENTS OF TRANSIENT RESPONSE V(3)

DC COMPONENT = -4.608223E-01
```

HARMONIC NO	FREQUENCY (HZ)	FOURIER COMPONENT	NORMALIZED COMPONENT	PHASE (DEG)	NORMALIZED PHASE (DEG)
1	5.000E+02	9.608E+00	1.000E+00	4.678E+01	0.000E+00
2	1.000E+03	6.216E-02	6.470E-03	-1.736E+02	-2.204E+02
3	1.500E+03	4.046E+00	4.211E-01	1.655E+01	-3.023E+01
4	2.000E+03	3.145E-02	3.273E-03	-1.758E+02	-2.226E+02
5	2.500E+03	2.484E+00	2.585E-01	8.511E+00	-3.827E+01
6	3.000E+03	2.099E-02	2.184E-03	-1.761E+02	-2.229E+02
7	3.500E+03	1.784E+00	1.857E-01	4.470E+00	-4.231E+01
8	4.000E+03	1.573E-02	1.638E-03	-1.759E+02	-2.227E+02
9	4.500E+03	1.390E+00	1.447E-01	1.838E+00	-4.495E+01

```
TOTAL HARMONIC DISTORTION = 5.473805E+01 PERCENT
```

The latter value agrees with that obtained from PSPICE simulation shown in Table 2.19.

The output waveform and its spectrum are shown in Figure 2.25.

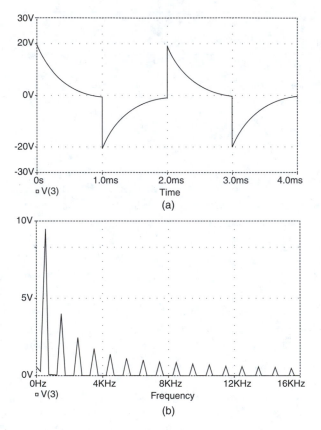

FIGURE 2.25
(a) Output waveform and (b) output spectra.

Bibliography

1. Al-Hashimi, Bashir, *The Art of Simulation Using PSPICE, Analog, and Digital*, CRC Press, Boca Raton, FL, 1994.
2. Al-Hashimi, Bashir, Behavioral Model Emulates Universal Filter, *EDN Magazine*, p. 124, Feb. 4, 1992.
3. Antognetti, Paolo and Massobrio, Giuseppe, *Semiconductor Device Modeling with SPICE*, McGraw-Hill, New York, 1993.
4. Brown, William L. and Szeto, Andrew Y. J., Verifying Spice Results with Hand Calculations: Handling Common Discrepancies, *IEEE Trans. Education*, 37(4), 358–368, 1994.
5. Connelly, J. Alvin and Choi, Pyung, *Macromodeling with SPICE*, Prentice-Hall, Englewood Cliffs, NJ, 1992.

6. Conrad, William R., Solving Laplace Transform Equation Using PSPICE, *Computers in Education Journal*, 5(1), 35–37, 1995.

7. Distler, R.J, Monte Carlo Analysis of System Tolerance, *IEEE Trans. Education*, 20, 98–101, May 1997.

8. Ellis, George, Use SPICE to Analyze Component Variations in Circuit Design, *EDN*, 109–114, April 1993.

9. Eslami, Mansour and Marleau, Richard S., Theory of Sensitivity of Network: A Tutorial, *IEEE Trans. Education*, 32(3), 319–334, August 1989.

10. Fenical, L.H., *PSPICE: A Tutorial*, Prentice-Hall, Englewood Cliffs, NJ, 1992.

11. Hamann, J.C., Pierre, J.W., Legowski, S.F., and Long, F.M., Using Monto Carlo Simulations to Introduce Tolerance Design to Undergraduates, *IEEE Trans. Education*, 42(1), 1–14, Feb. 1999.

12. Hart, Daniel W., Introducing Fourier Series Using PSPICE Computer Simulation, *Computers in Education, Division of ASEE*, III(2), 46–51, April–June 1993.

13. Kavanaugh, Micheal F., Including the Effects of Component Tolerances in the Teaching of Courses in Introductory Circuit Design, *IEEE Trans. Education*, 38(4), 361–364, Nov. 1995.

14. Kielkowski, Ron M., Inside SPICE, Overcoming the Obstacles of Circuit Simulation, McGraw-Hill, New York, 1994.

15. Lamey, Robert, *The Illustrated Guide to PSPICE*, Delmar Publishers, Albany, NY, 1995.

16. Monssen, Franz, *PSPICE with Circuit Analysis*, MacMillan, New York, 1992.

17. Nilsson, James W. and Riedel, Susan A., *Introduction to PSPICE*, Addison-Wesley, Reading, MA, 1993.

18. OrCAD PSPICE A/D, Users Guide, Nov. 1998.

19. Prigozy, Stephen, Novel Applications of PSPICE in Engineering, *IEEE Trans. Education*, 32(1), 35–38, Feb. 1989.

20. Rashid, Mohammad H., *SPICE for Power Electronics and Electric Power*, Prentice-Hall, Englewood Cliffs, NJ, 1993.

21. Rashid, Mohammad H., *SPICE for Circuits and Electronics Using PSPICE*, Prentice-Hall, Englewood Cliffs, NJ, 1990.

22. Roberts, Gordon W. and Sedra, Adel S., *SPICE for Microelectronic Circuits*, Saunders College Publishing, Fort Worth, TX, 1992.

23. Spence, Robert and Soin, Randeep S., *Tolerance Design of Electronic Circuits*, Imperial College Press, River Edge, NY, 1997.

24. Thorpe, Thomas W., *Computerized Circuit Analysis with SPICE*, John Wiley & Sons, New York, 1991.

25. Tuinenga, Paul W., *SPICE, A Guide to Circuit Simulations and Analysis Using PSPICE*, Prentice-Hall, Englewood Cliffs, NJ, 1995.

26. Vladimirescu, Andrei, *The SPICE Book*, John Wiley & Sons, New York, 1994.

27. Wyatt, Micheal A., Model Ferrite Beads in SPICE, *Electronic Design*, 76, Oct. 15, 1992.

Problems

2.1 For the circuit shown in Figure P2.1 the resistance R = 10 KΩ. (a) Find the change in the center frequency as C varies from 20 pF to 40 pF. (b) Plot the frequency response for the above values of C.

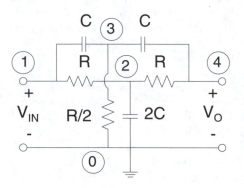

FIGURE P2.1
Twin-T network.

2.2 For the notched filter circuit shown in Figure P2.1, R1 = 10 KΩ and C = 20 pF. Assuming that the capacitors and resistors are sensitive to temperature. Find the change in the notch frequency if the TC1 of the resistors and capacitors is 1E–5. TC2 of resistors and capacitors is assumed to be zero.

2.3 For the voltage regulator circuit shown in Figure P2.3, VS = 20 V, RS = 300 Ω, RL = 4 KΩ, and D1 is D1N4742. Find the output voltage as a function of temperature as the latter varies from 25°C to 125°C.

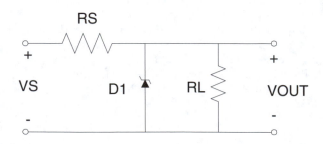

FIGURE P2.3
Voltage regulator.

2.4 For the voltage multiplier circuit shown in Figure 2.12(a), the waveform shown in Figure P2.4 is connected to the two input of the multiplier. Find the output voltage. Assume that K = 0.5.

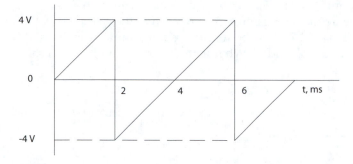

FIGURE P2.4
Input waveform.

2.5 An amplitude modulated wave is given as $s(t) = A_c \cos(2\pi f_c t + k_a m(t))$. If $m(t) = \cos(2\pi f_m t)$, $f_m = 10^4$ Hz, $f_c = 10^6$ Hz, $k_a = 0.5$, and $A_c = 10$ V, sketch $m(t)$ and $s(t)$ using the analog behavioral model technique.

2.6 A Zener diode in the voltage regulator circuit of Figure P2.6 has the corresponding current and voltage relationship shown in Table P2.6. R1 = R3 = R4 = 5 KΩ and R2 = 25 KΩ. Find the output voltage of the voltage regulator circuit when the input changes from 20 V to 25 V.

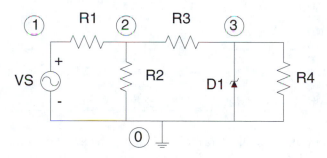

FIGURE P2.6
Voltage regulator circuit.

TABLE P2.6

Zener Diode Characteristics

Reverse Voltages (V)	Reverse Current (A)
1	1.0 E–11
3	1.0E–11
4	1.0E–10
5	1.0E–9
6	1.0E–7
7	1.0E–6
7.5	2E–6
7.7	15.0E–4
7.9	44.5E–4

2.7 Figure P2.7 shows a cascaded system with subsystems A, B, and C. The frequency response data of each of the cascaded system is shown in Table P2.7. Plot the frequency magnitude responses at the output of circuits A and C. What is the bandwidth of the complete system?

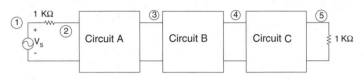

FIGURE P2.7
Cascaded circuit.

TABLE P2.7

Frequency Response of Each Subsystem A, B, and C

Frequency (Hz)	Gain of Circuit A (dB)	Gain of Circuit B (dB)	Gain of Circuit C (dB)
1000	10.0	0.8	2
2000	9.5	1.2	2
3000	8.3	1.5	2
4000	6.5	2.0	2
5000	5.0	2.5	2
6000	4.0	3.0	2
7000	3.0	3.5	2
8000	2.5	5.0	2
9000	2.0	6.5	2
10000	1.5	8.3	2
11000	1.2	9.5	2
12000	0.8	10.0	2

2.8 In Example 2.14, calculate the rms value of the input periodic waveform. What is the percent total harmonic distortion of the input waveform?

2.9 A system has a Laplace transform given as

$$\frac{V_{out}}{V_{in}}(s) = \frac{s^2 + 2s + 3}{s^3 + 9s^2 + 15s + 25}$$

Assuming the input is a step voltage with an amplitude of 5 V, determine the response of the system.

2.10 The open-loop gain of a frequency compensated op amp is given as

$$\frac{V_{out}}{V_{in}}(s) = \frac{A_0}{1 + \dfrac{s}{(2\pi f_p)}}$$

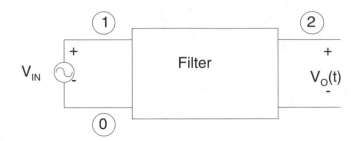

FIGURE P2.9
System with Laplace transform.

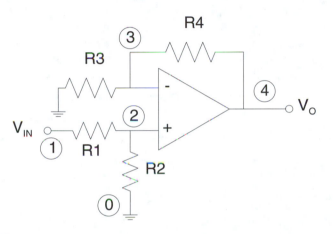

FIGURE P2.10
An operational amplifier circuit.

For the circuit shown in Figure P2.10, R1 = 1 KΩ, R2 = 2 KΩ, R3 = 4 KΩ, and R4 = 9 KΩ. If the op amp is frequency compensated with $A_0 = 10^5$, $f_p = 10$Hz, plot the frequency response of V_{out}/V_{in}.

2.11 For the Sallen-Key filter circuit shown in Figure P2.11, R1 = R2 = R3 = 10 KΩ, R4 = 40 KΩ, C1 = C2 = 0.02 μF, VCC = 15 V, and VEE = −15 V. If resistors and capacitors have 5% tolerance values, calculate the range of output voltage from the nominal value for 25, 50, 100 and 125 runs. Assume that a UA741 op amp is used, and the source has a frequency of 50 Hz and an amplitude of 1 V.

2.12 For the fifth-order low-pass filter, R1 = R2 = R3 = R4 = R5 = 1KΩ and C1 = C2 = C3 = C4 = C5 = 0.001 μF (Figure P2.12). (a) If the resistors and capacitors have tolerances of 5%, calculate the range of values of V_0 from its nominal value. Use 50 runs. (b) If the resistors and capacitors have tolerances of 10%, calculate the range of values of V_0 from its nominal values. Use 50 runs. The source has a frequency of 1 KHz and an amplitude of 1 V.

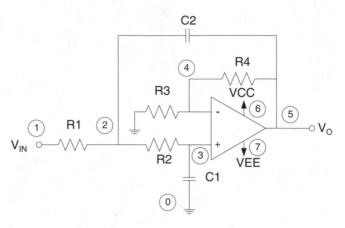

FIGURE P2.11
Sallen-Key filter circuit.

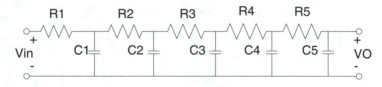

FIGURE P2.12
Fifth-order low-pass filter.

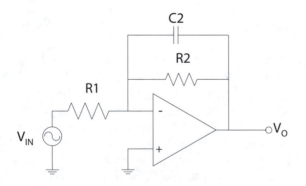

FIGURE P2.13
Miller integrator with dc gain.

2.13 The Miller integrator with gain at dc is shown in Figure P2.13. It has R1 = 10 KΩ, R2 = 100 KΩ and C2 = 0.01 μF. If the tolerance on the resistors is 5% and capacitors 5%, find the worst-case output voltage for frequencies between 10 Hz and 20 KHz. Assume that the op amp has an input resistance of $10^{10}\,\Omega$, zero output resistance, and an open loop gain of 10^8.

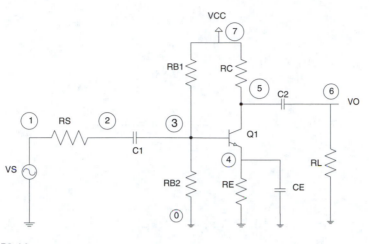

FIGURE P2.14
An amplifier circuit.

2.14 For the amplifier shown in Figure P2.14, RS = 150 Ω, RB2 = 20 KΩ, RB1 = 90 KΩ, RE = 2 KΩ, RC = 5 KΩ, RL = 10 KΩ, C1 = 2 μF, CE = 50 μF, C2 = 2 μF, and VCC = 15 V. If the resistors have a tolerance of 5% and the capacitors 5%, determine the sensitivity and the worst-case quiescent operating point of the transistor if the gain Bf of transistor is 100 with variation of 40% (uniform distribution). Assume that Q1 is a Q2N2222 transistor.

2.15 A full-wave rectifier is shown in Figure P2.15. RL = 10 K and D1, D2, D3, and D4 are D1N4009 diodes. If the input sinusoidal signal has a peak amplitude of 10 V and the frequency is 60 Hz, (a) find the Fourier series coefficients of the output load. (b) What is the rms value of the load voltage? (c) Calculate the power delivered to the load. (d) If a 10-μF capacitor is connected across the load resistor, use PROBE to display the frequency components of the output voltage.

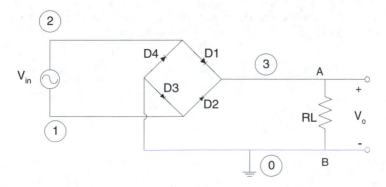

FIGURE P2.15
Full-wave rectifier circuit.

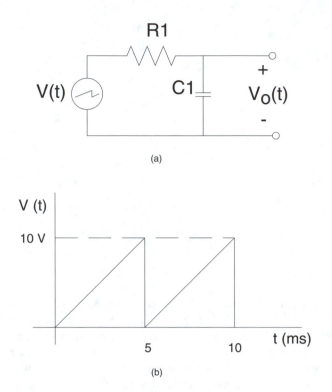

(a)

(b)

FIGURE P2.16
(a) RC network and (b) input sawtooth waveform.

2.16 A sawtooth waveform, shown in Figure P2.16(b), is applied at an input of the RC network shown in Figure P2.16(a). R1 = 2 KΩ and C1 = 5 µF. (a) Find the frequency components of the sawtooth waveform. (b) Determine the total harmonic distortion of the voltage across the capacitor. (c) What is the rms value of the voltage across the capacitor?

2.17 A sinusoidal signal $v_s(t)$ is applied at the input of an RLC circuit (Figure P2.17). R1 = 2 KΩ, L1 = 3 mH, and C1 = 0.1 µF. If $v_s(t) =$ 10 sin(2000 πt) volts, determine the frequency content of $v_o(t)$. What is the total harmonic distortion at the output?

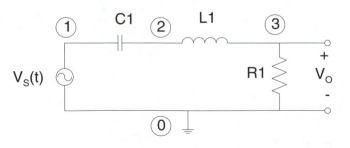

FIGURE P2.17
RLC circuit.

Section II

MATLAB Primer

3

MATLAB Fundamentals

MATLAB is numeric computation software for engineering and scientific calculations. The name MATLAB stands for MATRIX LABORATORY. MATLAB is primarily a tool for matrix computations. MATLAB has a rich set of plotting capabilities. The graphics are integrated in MATLAB. Because MATLAB is also a programming environment, a user can extend the functional capabilities to guarantee high accuracy. This chapter introduces the basic operations of MATLAB, the control statements, and the plotting functions.

3.1 MATLAB Basic Operations

When MATLAB is invoked, the command window will display the prompt >>. MATLAB is then ready for entering data or executing commands. To quit MATLAB, type the command **exit** or **quit**. MATLAB has online help. To see the list of MATLAB's help facility, type **help**. The **help** command followed by function names is used to obtain information on a specific MATLAB function. For example, to obtain information on the use of the fast Fourier transform function (fft), one can type the command **help fft**.

The basic data object in MATLAB is a rectangular numerical matrix with real or complex elements. Scalars are thought of as a 1-by-1 matrix. Vectors are considered as matrices with a row or column. MATLAB has no dimension statement or type declarations. Storage of data and variables is allocated automatically once the data and variables are used.

MATLAB statements are normally of the form

variable = expression

Expressions typed by the user are interpreted and immediately evaluted by the MATLAB system. If a MATLAB statement ends with a semicolon, MATLAB evaluates the statement but suppresses the display of the results. A matrix

$$A = \begin{bmatrix} 6 & 7 & 8 \\ 9 & 10 & 11 \\ 12 & 13 & 14 \end{bmatrix}$$

may be entered as follows:

A = [6 7 8; 9 10 11; 12 13 14];

Note that the matrix entries must be surrounded by brackets [ ] with row elements separated by blanks or by commas. A semicolon indicates the end of each row, with the exception of the last row. The matrix A can also be entered across three input lines as

```
A  = [6    7    8
         9   10   11
        12   13   14];
```

In this case, the carriage returns replace the semicolons. A row vector B with four elements

B = [30 40 60 90 71]

can be entered in MATLAB as

```
B  = [30   40   60   90   71];
```

or

```
B  = [30,   40,   60,   90,   71];
```

For readability, it is better to use spaces rather than commas between the elements. The row vector B can be turned into a column vector by transposition, which is obtained by typing

C = B'

The above results in

```
C   =
        30
        40
        60
        90
        71
```

TABLE 3.1

Some Basic MATLAB Commands

Command	Description
%	Comments. Everything appearing after the % command is not executed
demo	Access online demo programs
length	Length of matrix
clear	Clears the variables or functions from workspace
clc	Clears the command window during a work session
clg	Clears the graphic window
diary	Saves a session on disk, possibly for printing at a later date

Other ways of entering the column vector C are

```
C  =  [30
         40
         60
         90
         71]
```

or

```
C  =  [30;  40;  60;  90;  71]
```

MATLAB is case sensitive in naming variables, commands, and functions. Thus, **b** and **B** are *not* the same variables. If you do not want MATLAB to be case sensitive, you must use the command **casesen off**.

Table 3.1 shows additional MATLAB commands to get one started on MATLAB. Detailed descriptions and usages of the commands can be obtained from the MATLAB help facility or from MATLAB manuals.

3.2 Matrix Operations

The basic matrix operations are addition (+), subtraction (–), multiplication (*), and conjugate transpose(') of matrices. In addition to the above basic operations, MATLAB has two forms of matrix division: the left inverse operator \ or the right inverse operator /.

Matrices of the same dimension can be subtracted or added. Thus, if E and F are entered in MATLAB as

```
E  =  [21  25  30;  7  18  34;  70  16  8];
F  =  [1  7  3;  8  11  4;  2  11  13];
```

and

```
G  =  E  -  F
H  =  E  +  F
```

then matrices G and H will appear on the screen as

```
G  =

       20   18   27
       -1    7   30
       68    5   -5


H  =

       22   32   33
       15   29   38
       72   27   21
```

Matrix multiplication is defined provided the inner dimensions of the two operands are the same. Thus, if X is an n-by-m matrix and Y is an i-by-j matrix, X*Y is defined provided m is equal to i. Because matrices E and F are 3-by-3 matrices, the product

```
Q  =  E*F
```

results as

```
Q  =

       281   752   553
       219   621   535
       214   754   378
```

Any matrix can be multiplied by a scalar. For example,

```
2*Q
```

gives

```
ans  =

         562   1504   1106
         438   1242   1070
         428   1508    756
```

Note that if a variable name and the "=" sign are omitted, a variable name **ans** is automatically created.

TABLE 3.2

Some Common MATLAB Functions

Functions	Description
abs(x)	Calculates the absolute value of x
acos(x)	Determines $\cos^{-1}x$, with the results in radians
asin(x)	Determines $\sin^{-1}x$, with the results in radians
atan(x)	Calculates $\tan^{-1}x$, with the results in radians
atan2(x)	Obtains $\tan^{-1}(y/x)$ over all four quadrants of the circle; the results are in radians
cos(x)	Calculates $\cos(x)$, with x in radians
exp(x)	Computes e^x
log(x)	Determines the natural logarithm $\log_e(x)$
sin(x)	Calculates $\sin(x)$, with x in radians
sqrt(x)	Computes the square root of x
tan(x)	Calculates $\tan(x)$, with x in radians

Matrix division can either be the left division operator \ or the right division operator /. The right division **a/b**, for example, is algebraically equivalent to a/b, while the left division a\b is algebraically equivalent to b/a.

If $Z * I = V$ and Z is non-singular, the left division $Z\backslash V$ is equivalent to the MATLAB expression

$$I = inv(Z) * V$$

where **inv** is the MATLAB function for obtaining the inverse of a matrix.

The right division denoted by V/Z is equivalent to the MATLAB expression

$$I = V * inv(Z) \tag{3.1}$$

Apart from the function **inv**, there are additional MATLAB functions worth noting. They are given in Table 3.2.

The following example uses the **inv** function to determine the nodal voltages of a resistive circuit.

Example 3.1   Nodal Analysis of a Resistive Network

For the circuit shown in Figure 3.1, the resistances are in ohms (Ω). Write the nodal equations and solve for voltages V_1, V_2, and V_3.

Solution

Using Kirchoff's current law and assuming that currents leaving a node are positive, we have

For node 1:

$$\frac{V_1 - 20}{10} + \frac{V_1}{20} + \frac{V_1 - V_2}{25} = 0$$

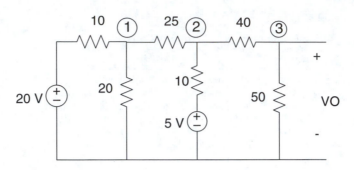

FIGURE 3.1
Resistive network.

Simplifying, we have

$$0.19V_1 - 0.04V_2 = 2 \tag{3.2}$$

For node 2:

$$\frac{V_2 - V_1}{25} + \frac{V_2 - 5}{10} + \frac{V_2 - V_3}{40} = 0$$

Simplifying, we have

$$-0.03V_1 + 0.165V_2 - 0.025V_3 = 0.5 \tag{3.3}$$

For node 3:

$$\frac{V_3}{50} + \frac{V_3 - V_2}{40} = 0$$

Simplifying, we have

$$-0.025V_2 + 0.045V_3 = 0 \tag{3.4}$$

In matrix form, Equations (3.2) through (3.4) become

$$\begin{bmatrix} 0.19 & -0.04 & 0 \\ -0.04 & 0.165 & -0.025 \\ 0 & -0.025 & 0.045 \end{bmatrix} \begin{bmatrix} V_1 \\ V_2 \\ V_3 \end{bmatrix} = \begin{bmatrix} 2 \\ 0.5 \\ 0 \end{bmatrix}$$

The MATLAB function **inv** is used to obtain the nodal voltages. MATLAB is used for solving the nodal voltages.

MATLAB script:

```
% This program computes the nodal voltages
% given the admittance matrix Y and current vector I
% Y is the admittance matrix
% I is the current vector
% Initialize the matrix y and vector I using YV=I
Y = [0.19 -0.04      0;
     -0.04  0.165  -0.025;
        0    -0.025   0.045];
% current is entered as a transpose of row vector
I = [2 0.5 0]';
fprintf('Nodal voltages V1, V2, and V3 are \n')
V = inv(Y)*I
```

We obtain the following results.
  Nodal voltages V1, V2, and V3 are

  V =

          11.8852
           6.4549
           3.5861

3.3   Array Operations

Array operations refer to element-by-element arithmetic operations. Preceding the linear algebraic matrix operations, * / \ ', by a period (.) indicates an array or element-by-element operation. Thus, the operators, .*, .\, ./, .^, represent element-by-element multiplication, left division, right division, and raising to the power, respectively. For addition and subtraction, the array and matrix operations are the same.

If K1 and L1 are matrices of the same dimensions, then K1.*L1 denotes an array whose elements are products of the corresponding elements of K1 and L1. Thus, if

  K1 = [1   7   4
        2   5   6];

  L1 = [11   12   14
         7   4   1];

then

```
M1  =  K1.*L1
```

results in

```
M1 =
     11   84   56
     14   20    6
```

An array operation for left and right division also involves element-by-element operation. The expressions **K1./L1** and **K1.\L1** give the quotient of element-by-element division of matrices K1 and L1. The statement

```
N1  =  K1./L1
```

gives the result

```
N1  =
          0.0909   0.5833   0.2857
          0.2857   1.2500   6.0000
```

And the statement

```
P1 = K1.\L1
```

gives

```
P1  =
          11.0000    1.7143    3.5000
           3.5000    0.8000    0.1667
```

Array exponentiation is denoted by .^. The general statement will be of the form:

```
q1 = r1.^s1
```

If r1 and s1 are matrices of the same dimensions, then the result q1 is also a matrix of the same dimensions. For example, if

```
r1  =  [4  3  7];
s1  =  [1  4  3];
```

then

```
q1  =  r1.^s1
```

gives the result

```
q1  =

      4    81   343
```

One of the operands can be scalar. For example,

```
q2  =  r1.^2
q3  =  (2).^s1
```

will give

```
q2  =

     16    9    49
```

and

```
q3  =

      2   16    8
```

Note that when one of the operands is scalar, the resulting matrix will have the same dimensions as the matrix operand.

3.4   Complex Numbers

MATLAB allows operations involving complex numbers. Complex numbers are entered using function **i** or **j**. For example, a number $z = 5 + j12$ can be entered in MATLAB as:

$$z = 5 + 12 * i$$

or

$$z = 5 + 12 * j$$

Also, a complex number $z1$

$$z1 = 4\sqrt{3} \exp\left[\left(\frac{\pi}{3}\right)j\right]$$

can be entered in MATLAB as:

$$z1 = 4 * sqrt(3) * \exp\left[\left(\frac{pi}{3}\right) * j\right]$$

It should be noted that when complex numbers are entered as matrix elements within brackets, one should avoid any blank spaces. For example, z2 = 5 +j12 is represented in MATLAB as:

z2 = 5+12*j

If spaces exist around the + sign, such as:

z3 = 5 + 12*j

MATLAB considers it as two separate numbers, and z2 will not be equal to z3.

If y is a complex matrix given as:

$$y = \begin{bmatrix} 1+j1 & 2-j2 \\ 3+j2 & 4+j3 \end{bmatrix}$$

then we can represent it in MATLAB as:

y = [1+j 2-2*j; 3+2*j 4+3*j]

which will produce the result

```
y  =
      1.0000  +  1.0000i      2.0000  -  2.0000i
      3.0000  +  2.0000i      4.0000  +  3.0000i
```

If the entries in a matrix are complex, then the "prime" (') operator pro-duces the conjugate transpose. Thus,

yp = y'

will produce

```
yp  =
      1.0000  -  1.0000i      3.0000  -  2.0000i
      2.0000  +  2.0000i      4.0000  -  3.0000i
```

TABLE 3.3

Some MATLAB Functions for Manipulating Complex Numbers

Function	Description
conj(Z)	Obtains the complex conjugate of a number Z. If Z = x + iy, then conj(Z) = x – iy
real(Z)	Returns the real part of the complex number Z
imag(Z)	Returns the imaginary part of the complex number Z
abs(Z)	Computes the magnitude of the complex number Z
angle(Z)	Calculates the angle of the complex number Z, determined from the expression `atan2(imag(Z), real(Z))`

For the unconjugate transpose of a complex matrix, we can use the point transpose (.') command. For example,

```
yt  =  y.'
```

will yield

```
yt=
     1.0000  +  1.0000i      3.0000  +  2.0000i
     2.0000  -  2.0000i      4.0000  +  3.0000i
```

There are several functions for manipulating complex numbers. Some of the functions are shown in Table 3.3.

Example 3.2   Input Impedance of Oscilloscope Probe

A simplified equivalent circuit of an oscilloscope probe for measuring low-frequency signals is shown in Figure 3.2. If R1 = 9 MΩ, R2 = 1 MΩ, C1 = 10 pF, and C2 = 100 pF, what is the input impedance at a sinusoidal frequency of 20 KHz?

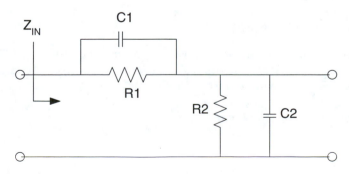

FIGURE 3.2
Simplified equivalent circuit of oscilloscope probe.

Solution

The input impedance is

$$Z_{IN} = \left[\frac{1}{j\omega C_1}\middle\|R_1\right] + \left[\frac{1}{j\omega C_2}\middle\|R_2\right]$$

$$= \frac{R_1}{1+j\omega C_1 R_1} + \frac{R_2}{1+j\omega C_2 R_2}$$

(3.5)

MATLAB is used to evaluate Z_{IN} for various values of frequency w.

The MATLAB script:

```
% ZIN is input impedance
c1 = 10e-12; c2=100e-12;
r1 = 9e+6;      r2=1.0e+6;
w = 2*pi*20.0e+3;
z1 = 1 + j*w*c1*r1;
z2 = 1 + j*w*c2*r2;
zin = (r1/z1) + (r2/z2);
zin
```

The solution obtained is

```
zin =
        7.6109E+004      -8.6868E+005i
```

3.5   The Colon Symbol

The colon symbol (:) is one of the most important operators in MATLAB. It can be used to (1) create vectors and matrices, (2) specify submatrices and vectors, and (3) perform iterations.

1. Creation of vectors and matrices
The statement

```
j1 = 1:8
```

will generate a row vector containing the numbers from 1 to 8 with unit increment. MATLAB produces the result

```
j1  =
        1  2  3  4  5  6  7  8
```

Non-unity, positive, or negative increments can be specified. For example, the statement

```
j2  =  4:-0.5:1
```

will yield the result

```
j2  =
        4.0000 3.5000 3.0000 2.5000 2.0000 1.5000 1.0000
```

The statement

```
j3  =  [(0:2:10);  (5:-0.2:4)]
```

will result in a 2-by-4 matrix

```
j3  =
        0          2.0000   4.0000   6.0000   8.0000   10.0000
        5.0000   4.8000   4.6000   4.4000   4.2000   4.0000
```

2. Specifying submatrices and vectors

Individual elements in a matrix can be referenced with subscripts inside parentheses. For example, j2(4) is the fourth element of vector j2. Also, for matrix j3, j3(2, 3) denotes the entry in the second row and third column. Using the colon as one of the subscripts denotes all of the corresponding row or column. For example, j3(:,4) is the fourth column elements of matrix j3. Thus, the statement

```
j5  =  j3(:,  4)
```

will give

```
j5  =
        6.0000
        4.4000
```

Also, the statement j3(2,:) is the second row of matrix j3. That is, the statement

```
j6  =  j3(2,:)
```

will result in

```
j6  =
        5.0000   4.8000   4.6000   4.4000   4.2000   4.0000
```

If the colon exists as the only subscript, such as j3(:), the latter denotes the elements of matrix j3 strung out in a long column vector. Thus, the statement

```
j7  =  j3(:)
```

will result in

```
j7  =
        0
        5.0000
        2.0000
        4.8000
        4.0000
        4.6000
        6.0000
        4.4000
        8.0000
        4.2000
       10.0000
        4.0000
```

3. Iterative uses of colon command

The iterative uses of the colon command are discussed in Section 3.6.

3.6   FOR Loops

"FOR" loops allow a statement or group of statements to be repeated a fixed number of times. The general form of a **for loop** is

> **for index = expression**
> **statement group C**
> **end**

The expression is a matrix and the statement group C is repeated as many times as the number of elements in the columns of the expression matrix.

The index takes on the elemental values in the matrix expression. Usually, the expression is something like

m:n or m:i:n

where **m** is the beginning value, **n** the ending value, and **i** is the increment.

Suppose we would like to find the cubes of all the integers starting from 1 to 50. We could use the following statement to solve the problem

```
sum = 0;
for i = 1:50
    sum = sum + i^3;
end
sum
```

For loops can be nested, and it is recommended that the loop be indented for readability. If we want to fill a 5-by-6 matrix, a, with an element value equal to unity, the following statements can be used to perform the operation:

```
%
n = 5; %number of rows
m = 6; %number of columns
for i = 1:n
    for j = 1:m
    a(i,j) = 1; %semicolon suppresses printing in the
loop
    end
end
a %display the result
%
```

It is important to note that each for statement group must end with the word **end**. The following example illustrates the use of the **for** loop.

Example 3.3   Frequency Response of a Notched Filter

A notched filter eliminates a small band of frequencies. It has a transfer function given as

$$H(s) = \frac{k_p\left(s^2 + \omega_0^2\right)}{s^2 + \left(\dfrac{\omega_0}{Q}\right)s + \omega_0^2} \tag{3.6}$$

If $k_p = 5$, $\omega_0 = 2\pi(5000)$ rads/s, $Q = 20$, calculate the values of $|H(s)|$ for frequencies between 4500 and 5500 Hz with increments of 50 Hz.

Solution

When $s = jw$, Equation (3.6) becomes

$$H(j\omega) = \frac{k_p\left[(j\omega)^2 + \omega_0^2\right]}{(j\omega)^2 + \left(\dfrac{\omega_0}{Q}\right)j\omega + \omega_0^2} = \frac{k_p\left[\omega_0^2 - \omega^2\right]}{\omega_0^2 - \omega^2 + j\dfrac{\omega_0}{Q}\omega} \qquad (3.7)$$

MATLAB is used to compute $H(jw)$ for various values of w.

MATLAB script:

```
% magnitude of H(j)
kp = 5;  Q = 20;   w0 = 2*pi*5000;
kj = w0/Q;
%
for i = 1:21
    w(i) = 2*pi*(4500 + 50*(i-1));
    wh(i) = w0^2 - w(i)^2;
    h(i) = kp*wh(i)/(wh(i) + j*kj*w(i));
h_mag(i) = abs(h(i));   % magnitude of transfer function
end
% print results
for k = 1:21
    w(k)/2*pi
    h_mag(k)
end
```

The results are given in the table below.

Frequency (Hz)	Magnitude
4.4413E+004	4.8654
4.4907E+004	4.8335
4.5400E+004	4.7898
4.5894E+004	4.7278
4.6387E+004	4.6363

(continued)

4.6881E+004	4.4949
4.7374E+004	4.2643
4.7868E+004	3.8651
4.8361E+004	3.1428
4.8855E+004	1.8650
4.9348E+004	0
4.9842E+004	1.8490
5.0335E+004	3.1047
5.0828E+004	3.8179
5.1322E+004	4.2166
5.1815E+004	4.4502
5.2309E+004	4.5953
5.2802E+004	4.6905
5.3296E+004	4.7558
5.3789E+004	4.8025
5.4283E+004	4.8369

From the above results, the notch frequency is 493.48 KHz.

3.7   IF Statements

IF statements use relational or logical operations to determine what steps to perform in the solution of a problem. The relational operators in MATLAB for comparing two matrices of equal size are shown in Table 3.4.

When any one of the above relational operators is used, a comparison is done between the pairs of corresponding elements. The result is a matrix of ones and zeroes, with **one** (1) representing **TRUE** and zero (0) representing **FALSE**. For example, if we have

```
ca  =  [1  7  3  8  3  6];
cb  =  [1  2  3  4  5  6];
ca  ==  cb
```

TABLE 3.4

Relational Operators

Relational Operator	Meaning
<	Less than
< =	Less than or equal
>	Greater than
> =	Greater than or equal
==	Equal
~ =	Not equal

TABLE 3.5

Logical Operators

Logical Operator	Meaning
&	and
!	or
~	not

The answer obtained is

```
ans  =
      1  0  1  0  0  1
```

The 1s indicate the elements in vectors ca and cb that are the same and 0s are the ones that are different.

There are three logical operators in MATLAB. These are shown in Table 3.5.

Logical operators work element-wise and are usually used on 0-1 matrices, such as those generated by relational operators. The & and ! operators compare two matrices of equal dimensions. If A and B are 0-1 matrices, then A&B is another 0-1 matrix with 1s representing TRUE and 0s FALSE. The NOT (~) operator is a unary operator. The expression ~C returns 1 where C is zero, and 0 when C is non-zero.

There are several variations of the **if** statement:

- Simple if statement
- Nested if statement
- If-else statement

The general form of the simple **if** statement is

if logical expression 1

　　　statement group G1

end

In the case of a simple **if** statement, if the logical expression 1 is true, the statement group G1 is executed. However, if the logical expression is false, the statement group G1 is bypassed and the program control jumps to the statement that follows the **end** statement.

The general form of a nested **if** statement is

if logical expression 1

　　　　　statement group G1

if logical expression 2

　　　statement group G2

    end

        statement group G3

    end

    statement group G4

The program control is such that if expression 1 is true, then statement groups G1 and G3 are executed. If the logical expression 2 is also true, the statement groups G1 and G2 will be executed before executing statement group G3. If logical expression 1 is false, we jump to statement group G4 without executing statement groups G1, G2, and G3.

    The **if-else** statement allows one to execute one set of statements if a logical expression is true and a different set of statements if the logical statement is false. The general form of the **if-else** statement is

    if logical expression 1

        statement group G1

    else

        statement group G2

    end

In the above program segment, statement group G1 is executed if logical expression 1 is true. However, if logical expression 1 is false, statement group G2 is executed.

    The **if-elseif** statement can be used to test various conditions before executing a set of statements. The general form of the **if-elseif** statement is

    if logical expression 1

        statement group G1

    elseif logical expression 2

        statement group G2

    elseif logical expression 3

        statement group G3

    elseif logical expression 4

        statement group G4

    end

A statement group is executed provided the true logical expression above is true. For example, if logical expression 1 is true, then statement group G1 is executed. If logical expression 1 is false and logical expression 2 is true, then statement group G2 will be executed. If logical expressions 1, 2, and 3 are false and logical expression 4 is true, then statement group G4 will be executed. If none of the logical expressions is true, then statement groups G1, G2, G3, and G4 will not be executed. Only three elseif statements are used in the above example. More elseif statements can be used if the application requires them. The following example illustrates the use of the **if** statement.

Example 3.4   Output Voltage of an Asymmetrical Limiter

The circuit shown in Figure 3.3 is a limiter. This circuit limits the output voltage to a specific value provided the input voltage exceeds or is lower than some threshold voltages. If the transfer function of the limiter is given as

$$v_0(t) = 3.0\,V \qquad for\ v_s(t) > 3.0\,V$$
$$= v_s(t) \qquad -4.0\,V \le v_s(t) \le 3.0V \qquad (3.8)$$
$$= -4.0V \qquad for\ v_s(t) < -4.0V$$

where we have assumed that the conducting diode has a 0.7-V drop, write a MATLAB program to obtain the output voltage from 0 to 24 seconds.

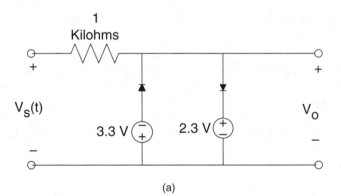

(a)

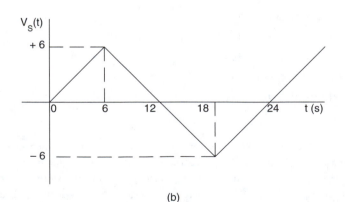

(b)

FIGURE 3.3
(a) Limiter circuit and (b) input voltage.

.
Solution

MATLAB script:

```
% vo is the output voltage
% vs is the input voltage
%
%Generate the triangular wave
for i = 1:25
    k = i-1;
    if i < = 7
        vs(i) = k;
    elseif i > = 7 & i < = 19
        vs(i) = 12 - k;
    else
        vs(i) = -24 + k;
    end
end
% Generate output voltage using if statement
for j = 1:25
    if vs(j) > = 3.0
        vo(j) = 3.0;
    elseif vs(j) < = -4.0
        vo(j) = -4.0;
    else
        vo(j) = vs(j);
    end
end
% print results
vs
vo
```

The results are

```
vs =

Columns 1 through 12
   0  1  2  3  4  5  6  5  4  3  2  1
```

(continued)

```
Columns 13 through 24

   0 -1 -2 -3 -4 -5 -6 -5 -4 -3 -2 -1

Column 25

   0

vo =

Columns 1 through 12

   0  1  2  3  3  3  3  3  3  3  2  1

Columns 13 through 24

   0 -1 -2 -3 -4 -4 -4 -4 -4 -3 -2 -1

Column 25

   0
```

Note that the output voltage is clipped at 3 V and –4 V.

3.8   Graph Functions

MATLAB has built-in functions that allow one to generate x-y, polar, contour, 3-D plots, and bar charts. MATLAB also allows one to give titles to graphs, label the x- and y-axes, and add grids to graphs. In addition, there are commands for controlling the screen and scaling. Table 3.6 shows a list of MATLAB built-in graph functions. One can use MATLAB's help facility to get more information on the graph functions.

3.8.1   X-Y Plots and Annotations

The plot command generates a linear x-y plot. There are three variations of the plot command:

1. **plot(x)**
2. **plot(x, y)**
3. **plot(x1, y1, x2, y2, x3, y3,...,xn, yn)**

If x is a vector, the command **plot(x)** produces a linear plot of the elements in the vector x as a function of the index of the elements in x. MATLAB will connect the points by straight lines. If x is a matrix, each column will be plotted as a separate curve on the same graph.

TABLE 3.5

Plotting Functions

Function	Description
axis	Freezes the axis limits
bar	Plots bar chart
contour	Performs contour plots
ginput	Puts cross-hair input from mouse
grid	Adds grid to a plot
gtext	Provides mouse-positioned text
histogram	Provides histogram bar graph
loglog	Does log vs. log plot
mesh	Performs 3-D mesh plot
meshdom	Provides domain for 3-D mesh plot
pause	Wait between plots
plot	Performs linear x-y plot
polar	Performs polar plot
semilogx	Does semilog x-y plot (x-logarithmic)
semilogy	Does semilog x-y plot (y-logarithmic)
stairs	Performs stair-step graph
text	Positions text at a specified location on graph
title	Used to put title on graph
xlabel	Labels x-axis
ylabel	Labels y-axis

If x and y are vectors of the same length, then the command **plot(x, y)** plots the element of x (x-axis) versus the elements of y (y-axis).

To plot multiple curves on a single graph, one can use the plot command with multiple arguments, such as **plot(x1, y1, x2, y2, x3, y3, …, xn, yn)**. The variables x1, y1, x2, y2, etc. are pairs of vectors. Each x-y pair is graphed, generating multiple lines on the plot. The above plot command allows vectors of different lengths to be displayed on the same graph. MATLAB automatically scales the plots. Also, the plot remains as the current plot until another plot is generated — in which case the old plot is erased.

When a graph is drawn, one can add a grid, a title, a label, and x- and y-axes to the graph. The commands for grid, title, x-axis label, and y-axis label are **grid** (grid lines), **title** (graph title), **xlabel** (x-axis label), and **ylabel** (y-axis label), respectively.

To write text on a graphic screen beginning at a point (x, y) on the graphic screen, one can use the **text(x,y,'text')** command. For example, the statement

```
text(2.0,1.5, 'transient analysis')
```

will write the text, transient analysis, beginning at point (2.0,1.5). Multiple text commands can be used. For example, the statements

```
plot(a1,b1,a2,b2)
text(x1,y1, 'voltage')
text(x2,y2, 'power')
```

TABLE 3.6

Print Types

Line-types	Indicators	Point Types	Indicators
Solid	-	Point	.
Dash	--	Plus	+
Dotted	:	Star	*
Dashdot	-.	Circle	o
		Xmark	x

will provide texts for two curves a1 versus b1 and a2 versus b2. The text will be at different locations on the screen provided $x1 - x2$ or $y1 \neq y2$.

If the default line-types used for graphing are not satisfactory, various symbols can be selected. For example,

plot(a1, b1, '*')

draws a curve, a1 versus b1, using star (*) symbols, while

plot(a1, b1, '*', a2, b2, '+')

uses a star (*) for the first curve and the plus (+) symbol for the second curve. Other print types are shown in Table 3.6.

For systems that support color, the color of the graph can be specified using the statement

plot(x, y, 'g')

implying plot x versus y using green color. Line and mark style can be added to color type using the command

plot(x, y, '+w')

The above statement implies plot x versus y using white + marks. Other colors that can be used are shown in Table 3.7.

The argument of the **plot** command can be complex. If z is a complex vector, then plot(z) is equivalent to plot(real(z), imag(z)). The following example shows the use of the **plot**, **title**, **xlabel**, and **ylabel** functions.

Example 3.5　Amplitude-Modulated Wave

A block diagram of an amplitude modulation of a communication system is shown in Figure 3.4. The double-sideband suppressed carrier $s(t)$ is given as

$$s(t) = m(t)c(t) \tag{3.9}$$

TABLE 3.7

Symbols for Color Used in Printing

Color	Symbol
Red	r
Green	g
Blue	b
White	w
Invisible	i

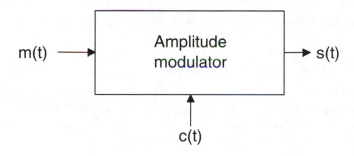

FIGURE 3.4
Block diagram of amplitude modulator.

If

$$m(t) = 2\cos(2000\pi t)\ V \tag{3.10}$$

$$c(t) = 10\cos\left(2\pi\left(10^6 t\right)\right) V \tag{3.11}$$

plot $s(t)$ from 0 to 60 μs.

Solution
MATLAB script:

```
% Amplitude modulated wave
% m(t) is the message signal
% c(t) is the carrier signal
% s(t) is the modulated wave
t = 0: 0.05e-6:3.0e-6;
k = length(t)
```

(continued)

```
for i = 1:k
    m(i) = 2*cos(2*pi*1000*t(i));
    c(i) = 10*cos(2*pi*1.0e+6*t(i));
    s(i) = m(i)*c(i);
end
plot(t, s, t, s,'o')
title('Amplitude Modulated Wave')
xlabel('Time in sec')
ylabel('Voltage, V')
```

The amplitude-modulated wave is shown in Figure 3.5.

3.8.2 Logarithmic and Plot3 Functions

Logarithmic and semi-logarithmic plots can be generated using the commands **loglog, semilogx,** and **semilogy**. The use of these **plot** commands is similar to those of the **plot** command discussed in the previous section. The descriptions of these commands are as follows:

> **loglog**(x,y): generates a plot of $\log_{10}(x)$ vs. $\log_{10}(y)$
> **semilogx**(x, y): generates a plot of $\log_{10}(x)$ vs. linear axis of y
> **semilogy**(x, y): generates a plot of linear axis of x vs. $\log_{10}(y)$

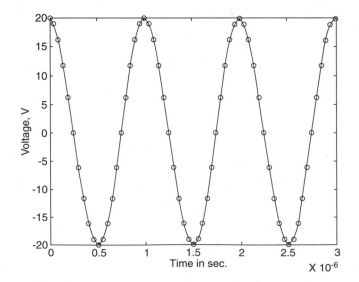

FIGURE 3.5
Amplitude-modulated wave.

TABLE 3.8

Frequency vs. Gain Data
of a High-Pass Network

Frequency (Hz)	Gain (dB)
30	15
50	20
100	25
200	35
500	50
1,000	65
4,000	85
6,000	90
10,000	92
50,000	97
100,000	99

It should be noted that since the logarithms of negative numbers and zero do not exist, the data to be plotted on the semi-log axes or log-log axes should not contain zero or negative values.

The **Plot3** function can be used to do three-dimensional line plots. The function is similar to the two-dimensional **plot** function. The **plot3** function supports the same line size, line style, and color options that are supported by the **plot** function. The simplest form of the **plot3** function is

　plot3(x, y, z)

where **x**, **y**, and **z** are equal-sized arrays containing the locations of the data points to be plotted.

The following example illustrates the use of the logarithmic plot.

Example 3.6 Magnitude Characteristics of a High-Pass Network

The gain vs. frequency of a high-pass network is shown in Table 3.8. Draw a graph of the gain vs. frequency.

Solution

A logarithmic scale is used for the frequency axis and a linear scale for gain. The MATLAB script is shown below.

MATLAB script:

```
% magnitude characteristics
% freq is the frequency values
freq = [30 50 100 200 500 1000 4000 6000 10000 50000
100000];
```

(continued)

```
% gain is the corresponding gain
gain = [15 20 25 35 50 65 85 90 92 97 99];
% use semilog to plot gain versus frequency
semilogx(freq, gain)
title('Characteristics of a High pass Network')
xlabel('Frequency in Hz')
ylabel('Gain in dB')
```

The magnitude characteristics of the network are shown in Figure 3.6.

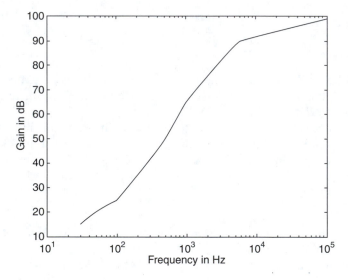

FIGURE 3.6
Gain vs. frequency of a high-pass network.

3.8.3 Subplot and Screen Control

MATLAB has two display windows: a command window and a graph window. The following commands can be used to select and clear these windows:

Command	Description
shg	Shows graph window
clc	Clears command window
clg	Clears graph window
home	Home command cursor

The graph window can be partitioned into multiple windows. The **subplot** command allows one to split the graph into subdivisions. Sub-windows can be arranged either top to bottom, or left to right. A four-window partition will have two sub-windows on top and two sub-windows on the bottom. The general form of the **subplot** command is

subplot(i j k)

The digits **i** and **j** specify that the graph window is to be split into an i-by-j grid of smaller windows, arranged in *i* rows and *j* columns. The digit *k* specifies the *k*th window for the current plot. The sub-windows are numbered from left to right, top to bottom.

TABLE 3.9

Numbering of Sub-windows for the subplot(3, 2, 4) Command

1	2
3	4 *(current figure)*
5	6

For example, the command **subplot (3 2 4)** creates six subplots in the current figure and makes subplot 4 the current plotting window. This is shown in Table 3.9.

The following example illustrates the use of the **subplot** command.

Example 3.7   Input and Output Voltages of a Schmitt Trigger Circuit

For the inverting Schmitt trigger circuit shown in Figure 3.7(a), R1 = 1 KΩ, R2 = 19 KΩ, and RS = 2 KΩ. The transfer characteristics of the circuit are shown in Figure 3.7(b). If the input voltage, $v_s(t)$ is a noisy signal given as

$$v_s(t) = 1.5\sin(2\pi f_0 t) + 0.8n(t) \tag{3.12}$$

where
f_0 = 500 Hz
$n(t)$ is a normally distributed white noise,

write a MATLAB program to find the output voltage. Plot both the input and output waveforms of the Schmitt trigger.

Solution

If $V_0(t)$ is the output voltage at time *t*, then the input and output voltages are related by the expressions

$$
\begin{aligned}
v_0(t) &= -10V &\quad &if \quad v_s(t) \geq 0.5V \\
&= -10V &\quad &if \quad -0.5V < v_s(t) < 0.5V \quad and \quad v_0(t-1) = -10V \\
&= +10V &\quad &if \quad -0.5V < v_s(t) < 0.5V \quad and \quad v_0(t-1) = +10V \\
&= +10V &\quad &if \quad v_s(t) \leq -0.5V
\end{aligned}
\tag{3.13}
$$

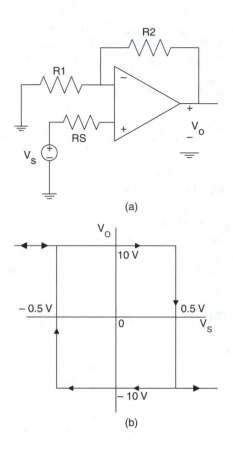

FIGURE 3.7
(a) Schmitt trigger circuit and (b) its transfer characteristics.

If statements will be used to execute relationships shown in Equation (3.13).

MATLAB script:

```
% vo is the output voltage
% vs is the input voltage
% Generate the sine voltage
t = 0.0:0.1e-4:5e-3;
fo = 500;   % frequency of sine wave
len = length(t)
for i = 1:len
    s(i) = 1.5*sin(2*pi*fo*t(i));
    % Generate a normally distributed white noise
    n(i) = 0.8*randn(1);
```

(continued)

```
    % generate the noisy signal
    vs(i) = s(i) + n(i);
end
% calculation of output voltage
%
len1 = len -1;
for i = 1:len1
    if  vs(i + 1) > = 0.5;
        vo(i+1) = -10;
    elseif  vs(i+1) > -0.5 & vs(i+1)< 0.5 & vo(i) == -10
        vo(i+1) = -10;
    elseif  vs(i+1) > -0.5 & vs(i+1)< 0.5 & vo(i) == +10
        vo(i+1) = 10;
    else
        vo(i+1) = +10;
    end
end
%
% Use subplots to plot vs and vo
    subplot (211), plot (t(1:40), vs(1:40))
    title ('Noisy time domain signal')
    subplot (212), plot (t(1:40), vo(1:40))
    title ('Output Voltage')
    xlabel ('Time in sec')
```

The input and output voltages are shown in Figure 3.8.

3.9   Input/Output Commands

MATLAB has commands for inputting information in the command window and outputting data. Examples of input/output commands are **echo, input, pause, keyboard, break, error, display, format**, and **fprintf**. Brief descriptions of these commands are shown in Table 3.10.

Break
The **break** command can be used to terminate the execution of **for** loops. If the **break** command exits in an innermost part of a nested loop, the **break**

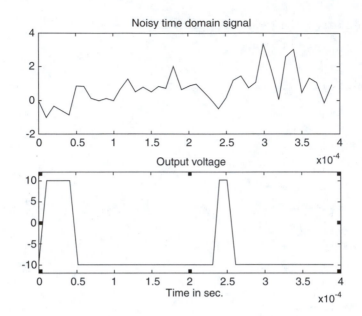

FIGURE 3.8
Input and output voltages.

TABLE 3.10

Some Input/Output Commands

Command	Description
break	Exits while or for loops
disp	Displays text or matrix
echo	Displays m-files during execution
error	Displays error messages
format	Displays output display to a particular format
fprintf	Displays text and matrices and specifies format for printing values
input	Allows user input
pause	Causes an m-file to stop executing; pressing any key causes interruptions of program execution

command will exit from that loop only. The **break** command is useful in exiting a loop when an error condition is detected.

Disp
The **disp** command displays a matrix without printing its name. It can also be used to display a text string. The general form of the **disp** command is

disp(x)
disp('text string')

disp(x) will display the matrix. Another way of displaying matrix x is to type its name. This is not always desirable because the display will start with a leading "x =". **Disp('text string')** will display the text string in quotes. For example, the MATLAB statement

disp('3-by-3 identity matrix')

will result in

3-by-3 identity matrix

Echo
The **echo** command can be used for debugging purposes. The **echo** command allows commands to be viewed as they execute. The **echo** command can be enabled or disabled:

echo on — enables the echoing of commands
echo off — disables the echoing of commands
echo — by itself toggles the echo statement

Error
The **error** command curve causes an error return from the m-files (discussed in Chapter 4) to the keyboard and displays a user-written message. The general form of the command is

Error('messages for display')

Consider the following MATLAB statements:

x=input('Enter age of student');
if x < 0
    error('wrong age was entered, try again')
end
x=input('Enter age of student')

For the above MATLAB statements, if the age is less than zero, the error message "Wrong age, try again" will be displayed and the user will be prompted for the correct age.

Format
The **format** commands control the format of an output. Table 3.11 shows some of the formats available in MATLAB.

By default, MATLAB displays numbers in "short" format (five significant digits). **Format compact** suppresses the line-feed that appears between

TABLE 3.11

Format Displays

Command	Meaning
format short	Five significant decimal digits
format long	Fifteen significant digits
format short e	Scientific notation with five significant digits
format long e	Scientific notation with fifteen significant digits
format hex	Hexadecimal
format +	+ printed if value is positive, – if negative; space is skipped if value is zero

matrix displays, thus allowing more lines of information to be seen on the screen. **Format loose** reverts to the less-compact display. **Format compact** and **format loose** do not affect the numeric format.

fprintf

The **fprintf** command can be used to print both text and matrix values. The format for printing the matrix can be specified and line-feed can also be specified. The general form of this command is

fprintf('text with format specification', matrices)

For example, the following statements, when executed,

```
res = 1.0e+6;
fprintf('The value of resistance is %7.3e Ohms\n', res)
```

will yield the output

```
The value of resistance is 1.000e+006 Ohms
```

The **format** specifier **%7.3e** is used to show where the matrix value should be printed in the text. 7.3e indicates that the resistance value should be printed with an exponential notation of seven digits, three (3) of which should be decimal digits. Other format specifiers include:

Format Specifier	Description
%c	Single character
%d	Decimal notation(signed)
%e	Exponential notation
%f	Fixed-point notation
%g	Signed decimal number in either %e or %f format, whichever is shorter

The text with format specification should end with \n to indicate the end of line. However, one can also use \n to get line-feeds, as represented by the following example:

```
r1 = 1500;
fprintf('resistance is \n%f Ohms \n',r1)
```

The output is:

```
resistance is
1500.000000 Ohms
```

Input

The **input** command displays a user-written text string on the screen, waits for an input from the keyboard, and assigns the number entered on the keyboard as the value of a variable. If the user enters a single number, it can be typed in. However, if the user enters an array, it must be enclosed in brackets. In either case, whatever is typed in will be stored in variable. For example, if one types the command

```
r = input('Please enter the three resistor values');
```

when the above command is executed, the text string `'Please, enter the three resistor values'` will be displayed on the terminal screen. The user can then type an expression such as:

```
[12 14 9]
```

The variable `r` will be assigned a vector [12 14 9]. If the user strikes the Return key without entering an input, an empty matrix will be assigned to `r`.

To return a string typed by a user as a text variable, the **input** command may take the form

```
x = Input('Enter string for prompt', 's')
```

Take for example, the command

```
x = input('What is the title of your graph', 's')
```

When executed, this command will echo on the screen, `'What is the title of your graph.'` The user can enter a string such as `'Voltage (mV) versus Current (mA).'`

One will get the echo

```
x =
     'Voltage (mV) versus Current (mA).'
```

Pause

The **pause** command stops the execution of m-files. The execution of the m-file resumes upon pressing any key. The general forms of the **pause** command are

pause

pause(n)

Pause stops the execution of m-files until a key is pressed. **Pause(n)** stops the execution of m-files for n seconds before continuing. The **pause** command can be used to temporarily stop m-files when plotting commands are encountered during program execution. If **pause** is not used, the graphics are momentarily visible.

The following example uses the MATLAB **input, fprint**, and **disp** commands.

Example 3.8   Equivalent Resistance of Series Connected Resistor

Write a MATLAB program that will accept values of resistors connected in series and find the equivalent resistance. The values of the resistors will be entered from the keyboard.

Solution

We use the MATLAB **input** command to accept the input of the elements, the **fprintf** command to output the result, and the **disp** command to display text string.

MATLAB script:

```
% input values of the resistors in input order
%
%
disp('Enter resistor values with spaces between them and
enclosed in brackets')
res = input('Enter resistor values')
num = length(res); % number of elements in array res
requiv = 0;
 for i = 1:num
    requiv = requiv +res(i);
end
%
fprintf('The Equivalent Resistance is %8.3e Ohms', requiv)
```

If you enter the values [2 3 7 9], you get the following result:

```
res =
     2   3   7   9
The Equivalent Resistance is 2.100e+001 Ohms
```

Bibliography

1. Attia, J.O., *Electronics and Circuit Analysis Using MATLAB*, CRC Press, Boca Raton, FL, 1999.
2. Biran, A. and Breiner, M., *MATLAB for Engineers*, Addison-Wesley, Reading, MA, 1995.
3. Chapman, S.J., *MATLAB Programming for Engineers*, Brook, Cole Thompson Learning, Pacific Grove, CA, 2000.
4. Derenzo, S.E., *Interfacing: A Laboratory Approach Using the Micrcomputer for Instrumentation, Data Analysis and Control*, Prentice-Hall, Englewood Cliffs, NJ, 1990.
5. Etter, D.M., *Engineering Problem Solving with MATLAB*, 2nd edition, Prentice-Hall, Upper Saddle River, NJ, 1997.
6. Etter, D.M., Kuncicky, D.C., and Hull, D., *Introduction to MATLAB 6*, Prentice-Hall, Upper Saddle River, NJ, 2002.
7. Gottling, J.G., *Matrix Analysis of Circuits Using MATLAB*, Prentice-Hall, Englewood Cliffs, NJ, 1995.
8. Rashid, M.H., *Microelectronic Circuits*, PWS Publishing, Boston, MA, 1999.
9. Sedra, A.S. and Smith, K.C., *Microelectronic Circuits*, 4th edition, Oxford University Press, New York, 1998.
10. Sigmor, K., MATLAB Primer, 4th edition, CRC Press, Boca Raton, FL, 1998.
11. Using MATLAB, The Language of Technical Computing, Computation, Visualization, and Programming, Version 6, MathWorks, Inc., 2000.

Problems

3.1 Find the nodal voltages V1, V2, V3, and V4 of Figure P3.1. The resistances are in ohms (Ω).

3.2 For the network shown in Figure P3.2, find nodal voltages, V1, V2, V3, and V4. The resistances are in ohms (Ω).

3.3 Find the current IO in Figure P3.3. The resistances are in ohms (Ω).

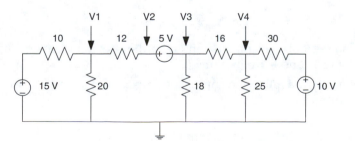

FIGURE P3.1
Circuit for Problem 3.1.

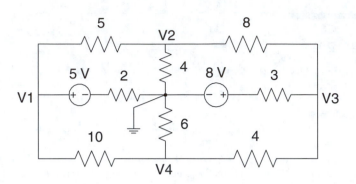

FIGURE P3.2
Circuit for Problem 3.2.

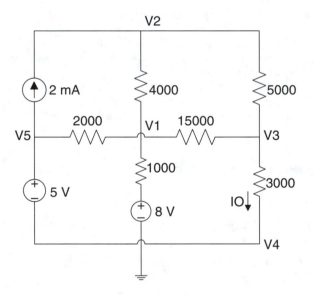

FIGURE P3.3
Circuit for Problem 3.3.

3.4 Find the loop currents I_1, I_2, and I_3 for the ladder network in Figure P3.4. The resistances are in ohms (Ω).

3.5 Simplify the following complex numbers and express them in rectangular and polar form.

(a) $za = 18 + j12 + \dfrac{(20 + j40)(5 - j15)}{25 + j25}$

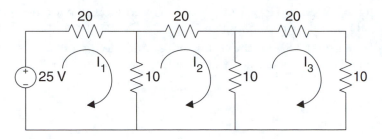

FIGURE P3.4
Ladder network.

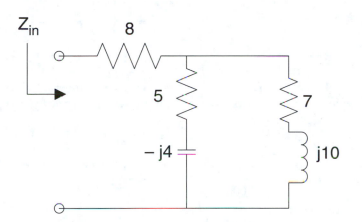

FIGURE P3.6
Circuit for Problem 3.6.

(b) $zb = \dfrac{10(-5+j13)(4+j4)}{(1+j2)(2+j5)(-5+j3)}$

(c) $zc = 0.2 + j7 + 4.7e^{j0.5} + (2+j3)e^{-j0.6\pi}$

3.6 Find the input impedance of the circuit shown in Figure P3.6. The impedances are in ohms (Ω).

3.7 The closed-loop gain G of an operational amplifier with a finite open-loop gain of A is given as

$$G = \dfrac{-\dfrac{R_2}{R_1}}{1+\left(\dfrac{1+\dfrac{R_2}{R_2}}{A}\right)}$$

If $R_2 = 20$ KΩ and $R_1 = 1$ KΩ, find the closed-loop gain for the following values of the open-loop gain: 10^2, 10^3, 10^4, 10^5, 10^6, and 10^7.

3.8 For Figure P3.8, find the equivalent admittance (in polar form) for the following frequencies: 1 KHz, 4 KHz, 7 KHz, and 10 KHz.

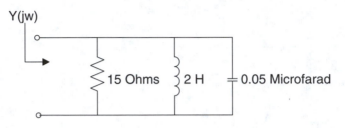

FIGURE P3.8
Parallel RLC circuit.

3.9 A limiter circuit shown in Figure P3.9. Assuming that a conducting diode has a 0.7-V drop, the relation between $i_s(t)$ and $v_s(t)$ is given as

$$i_s(t) = 0 \qquad\qquad \text{for } -6 \text{ V} < v_S(t) < 3.0 \text{ V}$$

$$= \frac{\left(v_s(t)-3\right)}{2000} \qquad\qquad \text{for } v_S(t) > 3.0 \text{ V}$$

$$= \frac{\left(v_s(t)+6.0\right)}{1000} \qquad\qquad \text{for } v_S(t) < -6.0 \text{ V}$$

If $v_s(t)$ is a square wave with a peak value of 10 V, an average value of 0 V, and a period of 4 ms, plot the input current $i_s(t)$ for one period of the input voltage. The resistances are in ohms (Ω).

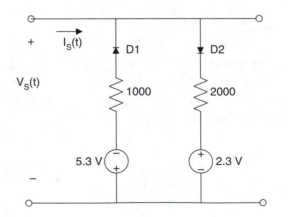

FIGURE 3.9
Limiter circuit.

3.10 The current flowing through the drain of a MOSFET is given as

$$i_{DS} = k_p \left(V_{GS} - V_T \right)^2 \quad A$$

If $V_T = 0.8$ V and $k_P = 4$ mA/V^2, plot i_{DS} for the following values of V_{GS}: 2, 2.5, and 3 V.

3.11 The equivalent impedance of the circuit, shown in Figure P3.11, is given as:

$$z(j\omega) = R + \frac{j\omega L}{1 - w^2 LC}$$

If $L = 1$ mH, $C = 10$ μF, and $R = 100$ Ω, plot the magnitude of the input impedance for w = 10, 100, 1000, 1.0E04, and 1.0E05 rads/s.

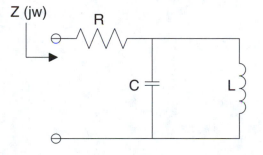

FIGURE P3.11
Circuit for Problem 3.11.

3.12 The voltage v_D and current i_D of a diode are related by the expression

$$i_D = I_s \exp\left(\frac{v_D}{n V_T} \right)$$

If $I_S = 10^{-16}$ A, $n = 1.5$, and $V_T = 26$ mV, plot i_D for diode voltages ranging from 0 to 0.65 V.

4

MATLAB Functions

In this chapter, MATLAB functions that will allow the user to process data from PSPICE simulations are discussed. The chapter begins with a discussion of m-files (script and function files). Some built-in MATLAB mathematical and statistical functions are introduced. In addition, four MATLAB functions — **diff**, **quad**, **quad8**, and **fzero** — are discussed. Furthermore, input/output functions are also covered. The chapter ends with methods of accessing results of PSPICE simulations by MATLAB.

4.1 M-Files

Normally, when single line commands are entered, MATLAB processes the commands immediately and displays the results. MATLAB is also capable of processing a sequence of commands that are stored in files with extension **m**. MATLAB files with extension m are called m-files, these are ASCII text files and are created with a text editor or word processor.

To list m-files in the current directory on your disk, you can use the MATLAB **what** command. The MATLAB **type** command can be used to show the contents of a specified file.

M-files can either be script or function. Script and function files contain a sequence of commands. However, function files take arguments and return values.

4.1.1 Script Files

Script files are especially useful for analysis and design problems that require long sequences of MATLAB commands. With script file written using a text editor or word processor, the file can be invoked by entering the name of the m-file without the extension. Statements in a script file operate globally on the workspace data.

Normally, when m-files are executing, the commands are not displayed on screen. The MATLAB **echo** command can be used to view m-files while they are executing. The MATLAB programs in Examples 3.1 through 3.8 are script files.

4.1.2   Function Files

Function files are m-files that are used to create new MATLAB functions. Variables defined and manipulated inside a function file are local to the function and do not operate globally on the workspace. However, arguments can be passed into and out of a function file.

The general form of a function file is

function variable(s) = function_name (arguments)
% help text in the usage of the function
%
.
.

The following is a summary of the rules for writing MATLAB m-file functions:

1. The word **function** appears as the first word in a function file. This is followed by an output argument, an equal sign, and the function name. The arguments to the function follow the function name and are enclosed within parentheses.
2. The information that follows the function, beginning with the % sign, shows how the function is used and what arguments are passed. This information is displayed if help is requested for the function name.
3. MATLAB can accept multiple input arguments, and multiple output arguments can be returned.
4. If a function is going to return more than one value, all the values should be returned as a vector in the function statement. For example,

$$\text{function[mean, variance]} = \text{data\_in}(x)$$

will return the mean and variance of a vector x. The mean and variance are computed with the function.
5. If a function has multiple input arguments, the **function** statement must list the input arguments. For example,

$$\text{function[mean, variance]} = \text{data}(x, n)$$

will return mean and variance of a vector x of length n.

The following example illustrates the usage of the m-file.

Example 4.1   Equivalent Resistance of Parallel-Connected Resistors

Write a function to find the equivalent resistance of parallel-connected resistors, R1, R2, R3,..., Rn.

Solution

MATLAB script:

```
function req=equiv_pr(r)
% equiv_pr is a function program for obtaining
%       the equivalent resistance of series
%       connected resistors
% usage: req=equiv_pr(r)
%       r is an input vector of length n
%
n=length(r);   % number of resistors
tmp=0.0;
for i=1:n
     tmp=tmp + 1/r(i);
end
req=1/tmp;
```

The MATLAB function shown above can be found in the function file **equiv_pr.m**.

Suppose we want to find the equivalent resistance of the parallel-connected resistors 2, 6, 7, 9, and 12 Ω. The following statements can be typed in the MATLAB command window to reference the function **equiv_pr**.

```
a = [2 6 7 9 12];
Rparall = equiv_pr(a)
```

The result obtained from MATLAB is

```
Rparall =
        0.9960
```

The equivalent resistance is 0.996 Ω.

4.2   Mathematical Functions

A partial list of mathematical functions available in MATLAB is shown in Table 4.1. A brief description of the various functions is also given. The following example illustrates the use of some of the mathematical functions.

TABLE 4.1

Common Mathematical Functions

Function Name	Explanation of Function
abs(x)	Absolute value or magnitude of complex number; calculates |x|
acos(x)	Inverse cosine; $\cos^{-1}(x)$; the results are in radians
angle(x)	Four-quadrant angle of a complex number; phase angle of complex number x in radians
asin(x)	Inverse sine, calculates $\sin^{-1}(x)$ with results in radians
atan(x)	Calculates $\tan^{-1}(x)$, with the results in radians
atan2(x,y)	Four-quadrant inverse; calculates $\tan^{-1}(y/x)$ over all four quadrants of the circle; the result is in radians in the range $-\pi$ to $+\pi$
ceil(x)	Round x to the nearest integer toward positive infinity; thus, ceil(4.2)=4; ceil(−3.3)=−3
conj(x)	Complex conjugate; i.e., x=3+j7; conj(x)=3−j7
cos(x)	Calculates cosine of x, with x in radians
exp(x)	Exponential; i.e., it calculates e^x
fix(x)	Rounds x to the nearest integer toward zero; fix(4.2)=4, fix(3.3)=3
floor(x)	Rounds x to the nearest integer toward minus infinity; floor(4.2)=4 and floor (3.3)=3
imag(x)	Complex imaginary part of x
log(x)	Natural logarithm: $\log_e(x)$
log10(x)	Common logarithm: $\log_{10}(x)$
real(x)	Real part of complex number x
rem(x,y)	Remainder after division of (x/y)
round(x)	Round toward nearest integer
sin(x)	Sine of x, with x in radians
sqrt(x)	Square root of x
tan(x)	Tangent of x

Example 4.2   Generation of a Full-Wave Rectifier Waveform

A full-wave rectifier waveform can be generated by passing a signal through an absolute value detector, whose block diagram is shown in Figure 4.1. If the input signal is $x(t) = 10\sin(120\pi t)$, and $y(t) = |x(t)|$, write a MATLAB program to plot x(t) and y(t).

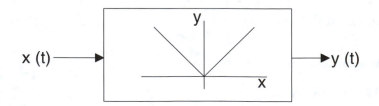

FIGURE 4.1
Block diagram of an absolute value circuit.

Solution

MATLAB script:

```
% x(t) in the input
% y is the output
period = 1/60;
period2 = 2*period;
inc = period/100;
npts = period2/inc;
for i = 1:npts
    t(i) = (i-1)*inc;
    x(i) = 10*sin(120*pi*t(i));
    y(i) = abs(x(i));
end
% plot x and y
subplot(211), plot(t,x)
ylabel('Voltage,V')
title('Input signal x(t)')
subplot(212), plot(t,y)
ylabel('Voltage,V')
xlabel('Time in seconds')
title('Output Signal y(t)')
```

The plots are shown in Figure 4.2.

From Figure 4.2, the output signal is a rectified version of the input waveform.

4.3   Data Analysis Functions

In MATLAB, data analyses are performed on column-oriented matrices. Different variables are stored in the individual column cells, and each row represents different observation of each variable. A data consisting of ten samples in four variables would be stored in a matrix of size 10-by-4. Functions act on the elements in the column. Table 4.2 gives a brief description of various MATLAB functions for performing data analysis. The following example illustrates the use of some of the data analysis functions.

Example 4.3   Statistics of Resistors

Resistances of three bins containing 1-KΩ, 10-KΩ, and 50-KΩ resistors, respectively, were measured using a multimeter. Ten resistors selected from

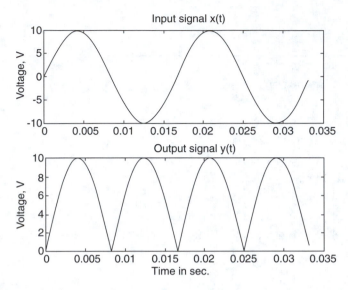

FIGURE 4.2
(Upper) Input sine waveform x(t) and (Lower) output waveform y(t).

the three bins have the values shown in Table 4.3. For each resistor bin, determine the mean, median, and standard deviation.

Solution

The data is shown in Table 4.3 is stored as a 3-by-10 matrix y.

MATLAB script:

```
% This program computes the mean, median, and standard
% deviation of resistors in bins
% the data is stored in matrix y
y = [1050 10250 50211;
992 9850 52500;
1021 9850 52500;
980 9752 53700;
1070 10102 48800;
940 9920 51650;
1005 10711 49220;
998 9520 54170;
1021 10550 46840;
987 9870 51100];
```

(continued)

```
%
% Calculate the mean
mean_r = mean(y);
% Calculate the median
median_r = median(y);
% Calculate the standard deviation
std_r = std(y);
% Print out the results
fprintf('Statistics of Resistor Bins\n\n')
fprintf('Mean of 1K,10K, 50K
bins,respectively:%7.3e,%7.3e, %7.3e \n', mean_r)
fprintf('Median of 1K, 10K, 50K bins: respectively:
%7.3e, %7.3e, %7.3e\n', median_r)
fprintf('Standard Deviation of 1K, 10K, 50K bins,
respectively: %7.8e, %7.8e, %7.8e \n', std_r)
```

TABLE 4.2

Data Analysis Functions

Function	Description
corrcoef(x)	Determines correlation coefficients
cov(x)	Obtains covariance matrix
cross(x, y)	Determines the cross product of vectors x and y
cumprod(x)	Finds a vector of the same size as x containing the cumulative products of the values from x; if x is a matrix, this function returns a matrix the same size as x containing cumulative products of values from the columns of x
cumsum(x)	Obtains a vector of the same size as x containing the cumulative sums of values from x; if x is a matrix, this function returns a matrix the same size as x containing cumulative values from the columns of x
diff(x)	Computes the differences between elements of an array x; it approximates derivatives; this function is discussed in detail in Section 4.3
dot(x, y)	Determines the dot product of vectors x and y
hist(x)	Draws the histogram or the bar chart of x
max(x)	Obtains the largest value of x; if x is a matrix, this function returns a row vector containing the maximum elements of each column
[y, k] = max(x)	Obtains the maximum value of x and the corresponding locations (indices) of the first maximum value for each column of x
mean(x)	Determines the mean or the average value of the elements in the vector; if x is a matrix, this function returns a row vector that contains the mean value of each column
median(x)	Finds the median value of the elements in the vector x; if x is a matrix, this function returns a row vector containing the median value of each column

TABLE 4.2 (continued)

Data Analysis Functions

Function	Description
min(x)	Finds the smallest value of x; if x is a matrix, this funciton returns a row vector containing the minimum values from each column
[y, k] = min(x)	Obtains the smallest value of x and the corresponding locations (indices) of the first minimum value from each column of x
prod(x)	Calculates the product of the elements of x; if x is a matrix, this function returns a row vector that contains the product of each column
rand(n)	Generates random numbers; if n = 1, a single random number is returned; if n > 1, an n-by-n matrix of random numbers is generated; this function generates the random numbers uniformly distributed in the interval [0,1]
rand(m, n)	Generates an m-by-n matrix containing uniformly distributed random numbers between 0 and 1
rand('seed', n)	Sets the seed number of the random number generator to n; if rand is called repeatedly with the same seed number, the sequence of random numbers become the same
rand('seed')	Returns the current value of the "seed" values of the random number generator
rand(m,n)	Generates an m-by-n matrix containing random numbers uniformly distributed between 0 and 1
randn(n)	Produces an n-by-n matrix containing normally distributed (Gaussian) random numbers with a mean of 0 and variance of 1
randn(m,n)	Produces an m-by-n matrix containing normally distributed (Gaussian) random numbers with a mean of 0 and variance of 1. To convert Gaussian random number r_n with mean value of 0 and variance of 1 to a new Gaussian random number with mean of μ and standard deviation σ, we use the conversion formula: $$X = \sigma \cdot r_n + \mu$$ Thus, random number data with 200 values, Gaussian distributed with mean value of 4 and standard deviation of 2 can be generated with the equation: $$data\_g = 2.randn(1, 200) + 4$$
sort(x)	Sorts the rows of a matrix in ascending order
std(x)	Calculates and returns the standard deviation of x if it is a one-dimensional array; if x is a matrix, a row vector containing the standard deviation of each column is computed and returned
sum(x)	Calculates and returns the sum of the elements in x; if x is a matrix, this function calculates and returns a row vector that contains the sum of each column
trapz(x,y)	Trapezoidal integration of the function y=f(x); a detailed discussion of this function is given in Section 4.5

TABLE 4.3

Resistances in 1-KΩ, 10-KΩ, and 50-KΩ Resistor Bins

Number	1-KΩ Resistor Bin	10-KΩ Resistor Bin	50-KΩ Resistor Bin
1	1050	10250	50211
2	992	9850	52500
3	1021	10460	47270
4	980	9752	53700
5	1070	10102	48800
6	940	9920	51650
7	1005	10711	49220
8	998	9520	54170
9	1021	10550	46840
10	987	9870	51100

The results are

Statistics of Resistor Bins

Mean of 1K,10K, 50K bins, respectively:1.006e+003,1.004e+004, 5.107e+004

Median of 1K, 10K, 50K bins: respectively:1.002e+003, 9.895e+003, 5.138e+004

Standard Deviation of 1K, 10K, 50K bins, respectively:3.67187509e+001, 3.69243446e+002, 2.31325103e+003

Before doing the next example, let us discuss the MATLAB function **freqs.** This function is used to obtain the frequency response of a transfer function. The general form is

$$H(s) = freqs(num, den, range) \tag{4.1}$$

where

$$H(s) = \frac{b_m s^m + b_{m-1} s^{m-1} + \ldots + b_1 s + b_0}{a_n s^n + a_{n-1} s^{n-1} + \ldots + a_1 s + a_0} \tag{4.2}$$

$$sum = b_m \, b_{m-1} \ldots b_1 \, b_0 \tag{4.3}$$

$$den = a_n \, a_{n-1} \ldots a_1 \, a_0$$

$range$ = a range of frequencies $\tag{4.4}$

$H(s)$ = frequency response (in complex form)

Example 4.4   Center Frequency of Band-Reject Filter

The transfer function of a band-reject filter is given as

$$H(s) = \frac{s^2 + 9.859 \times 10^8}{s^2 + 3140s + 9.859 \times 10^8} \tag{4.5}$$

Find the center frequency.

Solution

MATLAB script:

```
% numerator and denominator polynomial

num = [1 0 9.859e8];
den = [1 3.14e3 9.859e8];
w = logspace(-3,5,5000);
hs = freqs(num, den, w);   % finds frequency
f = w/(2*pi);   %finds frequency from rad/s to Hz
mag = 20*log10(abs(hs));   %magnitude of hs
% find minimum value of magnitude and its index
[mag_m floc] = min(mag);
% minimum frequency
fmin = f(floc);
%print results
fprintf('Minimum Magnitude (dB) is %8.4e\n', mag_m)
fprintf('Minimum frequency is %8.4e\n', fmin)
plot(f,mag)
```

The results obtained are

Minimum Magnitude (dB) is –3.1424E+001
Minimum frequency is 5.0040E+003

The center frequency is 5.004E+003 Hz.

4.4   Derivative Function (DIFF)

If *f* is a row or column vector

$$f = \left[f(1) \quad f(2) \ldots f(n) \right] \tag{4.6}$$

then the **diff(f)** function returns a vector containing the difference between adjacent elements; that is,

$$diff(f) = [f(2) - f(1), f(3) - f(2) \ldots f(n) - f(n-1)]$$
(4.7)

The output vector **diff(f)** will be one element less than the input vector f.

Numerical differentiation can be obtained using the backward difference expression

$$f'(x_n) = \frac{f(x_n) - f(x_{n-1})}{x_n - x_{n-1}}$$
(4.8)

or the forward difference

$$f'(x_n) = \frac{f(x_{n+1}) - f(x_n)}{x_{n+1} - x_n}$$
(4.9)

The derivative of $f(x)$ can be obtained using the MATLAB **diff** function as

$$f'(x) \cong \frac{diff(f)}{diff(x)}$$
(4.10)

The MATLAB function **diff** is used in the following example.

Example 4.5   Differentiator Circuit with Noisy Input Signal

An operational amplifier differentiator has input and output voltages related by the expression

$$V_0(t) = -k \frac{d}{dt} V_{IN}(t), \quad k = 0.0001$$
(4.11)

If the input voltage is given as

$$V_{IN}(t) = \sin(2\pi f_0 t) + 0.2n(t)$$

where

f_0 = 500 Hz

$n(t)$ = normally distributed white noise

Sketch $V_0(t)$ and $V_{IN}(t)$ using the **subplot** command.

Solution

MATLAB script:

```
% Differentiator circuit with noisy input
%
% generate input signal
%
t = 0.0:5e-5:6e-3;
k = -0.0001;
f0 = 500;
m = length(t);
% generate sine wave portion of signal
for i = 1:m
    s(i) = sin(2*pi*f0*t(i));
    % generate a normally distributed white noise
    n(i) = 0.2*randn(1);
    % generate noisy signal
    vin(i) = s(i) + n(i);
end
Subplot(211), plot(t(1:100), vin(1:100))
Title ('Noisy Input Signal')
% derivative of input signal is calculated using
% backward difference
dvin = diff(vin)./diff(t);
% output voltage is calculated
vout = k * dvin;
% plot the output voltage
subplot(212), plot(t(2:101), vout(1:100))
title('Output Voltage of Differentiator')
xlabel('Time in s')
```

Figure 4.3 shows the input and output of a differentiator.

4.5   Integration Functions (quad, quad8, trap)

The **quad** function uses an adaptive recursive Simpson's rule. However, the **quad8** function uses an adaptive recursive Newton Cutes 8 panel rule. The

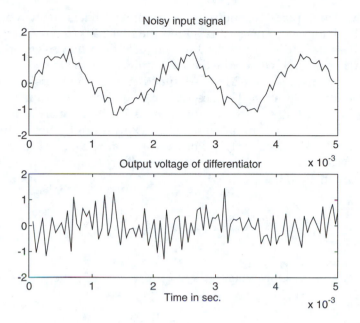

FIGURE 4.3
Input and output voltages of a differentiator.

quad8 function is better than **quad** at handling functions with "soft" singularities such as $\int \sqrt{x}\,dx$. Suppose we want to find S given as

$$S = \int_a^b funct(x)\,dx \qquad (4.12)$$

The general form of **quad** and **quad8** functions that can be used to find S is

    quad('funct', a, b, tol, trace)
    quad8('funct', a, b, tol, trace)

where
    funct is a MATLAB function name (in quotes) that returns a vector of values $f(x)$ for a given vector of input values x.
    a is the lower limit of integration.
    b is the upper limit of integration.
    tol (optional) is the tolerance limit set for stopping the iteration of the numerical integration. The iteration continues until the relative error is less than tol. The default value is 1.0E–3.
    trace (optional) allows the plot of a graph showing the process of numerical integration. If **trace** is non-zero, a graph is plotted. The default value is zero.

The **quad** and **quad8** functions use an argument that is an analytic expression of the integrand. This facility allows the functions (**quad** and **quad8**) to reduce the integration subinterval automatically until a given precision is attained. However, if we require the integration of a function whose analytic expression is unknown, the MATLAB function **trapz** can be used to perform the numerical integration. The description of the function **trapz** follows.

The MATLAB **trapz** function is used to obtain the numerical integration of a function (with or without an analytic expression) by use of the trapezoidal rules. If a function f(x) has known values at $x_1, x_2, ..., x_n$ given as $f(x_1)$, $f(x_2), ... f(x_n)$, respectively, then the trapezoidal rule approximates the area under the function; that is,

$$A \cong (x_2 - x_1)\left[\frac{f(x_1) + f(x_2)}{2}\right] + (x_3 - x_2)\left[\frac{f(x_2) + (x_3)}{2}\right] + ... \qquad (4.13)$$

For constant spacing where $x_2 - x_1 = x_3 - x_2 = ... = h$, the above equation reduces to:

$$A = h\left[\frac{1}{2}f(x_1) + f(x_2) + f(x_3) + ... f(x_{n-1}) + \frac{1}{2}f(x_n)\right] \qquad (4.14)$$

The error in the trapezoidal method of numerical integration reduces as the spacing h decreases.

The general form of the **trapz** function is

$$S2 = \mathbf{trapz}(x, y) \qquad (4.15)$$

where
   trapz(x, y) computes the integral of y with respect to x; x and y must be vectors of the same length.

Another form of the **trapz** function is

$$S2 = \mathbf{trapz}(Y) \qquad (4.16)$$

where
   trapz(Y) computes the trapezoidal integral of Y assuming unit spacing data points. If the spacing is different from 1, assuming it is **h**, then **trapz(Y)** should be multiplied by **h** to obtain the numerical integration.

That is, $S1 = (h)(S2) = (h).trapz(Y)$
   The following example shows the use of the **trapz** function.

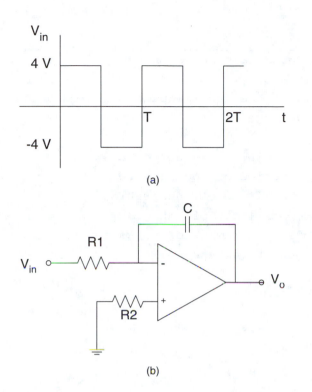

FIGURE 4.4
(a) Square-wave input and (b) op amp integrator.

Example 4.6   Integrator Circuit with a Square-Wave Input

A square wave, as shown in Figure 4.4(a), is applied at the input of an integrator, shown in Figure 4.4(b). R1 = R2 = 10 KΩ, C = 1 µF, and the period of the square wave is 2 ms. If the capacitor has zero initial voltage, (1) plot the output waveform and (2) calculate the root-mean-squared value of the output voltage.

Solution

For the op amp integrator, the output voltage V_0 is given as:

$$V_o(t) = -\frac{1}{RC} \int_0^t V_{IN}(\tau) d\tau \qquad (4.17)$$

Given the input voltage, the **trapz** function will be used to perform the numerical integration. The rms value of the output waveform is given as:

$$V_{0,rms} = \sqrt{\frac{1}{T_0} \int_0^{T_0} V_0^2 dt} \qquad (4.18)$$

MATLAB script:

```
% This program calculates the output voltage of
% an integrator
% In addition, we can calculate the rms voltage of the
% output voltage
%
R = 10e3; C = 1e-6; %values of R and C
T = 2e-3; %period of square wave
a = 0; %Lower limit of integration
b = T;   %Upper limit of integration
n = 0:0.005:1;   %Number of total data points
% Obtain output voltage
m = length(n);
% Generate time
for i = 1:m
        t(i) = T*i/m;
    if t(i) < 1e-3
        VX(i) = 4.0;
    else
        VX(i) = -4.0;
    end
    vo_int(i) = trapz(t(1:i), VX(1:i));
    vo(i) = -vo_int(i)/(R*C);% output voltage
    vo_sq(i) = vo(i)^2; % squared output voltage
end
%
plot(t(1:200), vo(1:200)),   % plot of vo
xlabel('Time in Sec')
Title('Output Voltage, V')
% Determine rms value of output
s = trapz(t(1:m), vo_sq(1:m));   % numerical integration
vo_rms = sqrt(s/b);   % rms value of output
%print out the result
fprintf('rms value of output is %7.3e\n', vo_rms)
```

The result obtained from MATLAB is:

   rms value of output is 2.275E-001

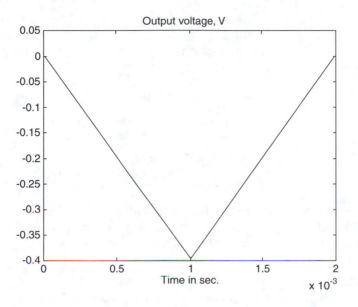

FIGURE 4.5
Output voltage of an op amp integrator.

The plot of the output voltage is shown in Figure 4.5.

4.6   Curve Fitting (polyfit, polyval)

The MATLAB **polyfit** function is used to compute the best fit of a set of data points to a polynomial with a specified degree. The general form of the function is

$$\textbf{poly\_xy} = \textbf{polyfit}(\textbf{x}, \textbf{y}, \textbf{n}) \qquad (4.19)$$

where
  x and y are the data points.
  n   is the nth degree polynomial that will fit the vectors x and y.
  poly_xy is a polynomial that fits the data in vector y to x in the least squares sense. **poly_xy** returns (n+1) coefficients in descending powers of x.

Thus, if the polynomial fit to vectors x and y is given as:

$$poly\_xy(x) = a_1 x^n + a_2 x^{n-1} + \dots a_m \qquad (4.20)$$

TABLE 4.4

Voltage and Current
of a Zener Diode

Diode Voltage v_D (V)	Current i_D (A)
−4.686	−1.187E−02
−4.694	−1.582E−02
−4.704	−2.376E−02
−4.708	−2.773E−02
−4.712	−3.170E−02
−4.715	−3.568E−02

The degree of the polynomial is **n** and the number of coefficients **m = n+1**. The coefficients $(a_1, a_2, ..., a_m)$ are returned by the MATLAB **polyfit** function. An application of the **polyfit** function is illustrated by the following example.

Example 4.7   Zener Diode Parameters from Data

A Zener diode at breakdown has the following corresponding voltage and current shown in Table 4.4. Plot the graph of current vs. voltage. Determine the dynamic resistance.

Solution

The dynamic resistance of the Zener diode is

$$r_D = \frac{\Delta v_D}{\Delta i_D} \qquad (4.21)$$

The plot of v_D vs. i_D is almost a straight line with the equation

$$i_D = m * v_D + I_0 \qquad (4.22)$$

The plot of v_D vs. i_D will have the slope given by $1/r_D$. MATLAB is used to plot the best fit linear model and to calculate resistance of the Zener diode.

MATLAB script:

```
%
% Diode parameters
vd = [-4.686 -4.694 -4.699 -4.704 ...
      -4.708 -4.712 -4.715];
id = [-1.187e-002 -1.582e-002 -1.978e-002 ...
```

(continued)

```
        -2.376e-002 -2.773e-002 -3.170e-002 ...
        -3.568e-002];
%
% coefficient
pfit = polyfit (vd, id, 1);
% Linear equation is y = m*x + b
b = pfit(2);
m = pfit(1);
ifit = m*vd + b;
% Calculate Is and n
rd = 1/m
% Plot v versus ln(i) and best fit linear model
plot (vd, ifit, 'b', vd, id, 'ob')
xlabel ('Voltage, V')
ylabel('Diode Current')
title('Best Fit Linear Model')
```

The results obtained from MATLAB are

rd =

   1.2135

The resistance of the Zener diode is 1.2135 Ω.

Figure 4.6 shows the best-fit linear model used to determine the resistance of the diode.

4.7 Polynomial Functions (roots, poly, polyval, and fzero)

4.7.1 Roots of Polynomials (roots, poly, polyval)

If $f(x)$ is a polynomial of the form

$$f(x) = C_0 x^n + C_1 x^{n-1} + \ldots + C_{n-1} x + C_n \tag{4.23}$$

$f(x)$ is of degree n and has exactly n roots. The n roots may have multiple **roots** or complex **roots**. If the coefficients of the polynomial (C_0, C_1, C_2, ..., C_m) of the polynomial are real, then the complex roots will occur as complex conjugate pairs.

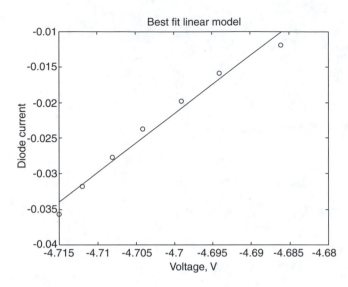

FIGURE 4.6
Current vs. voltage of a Zener diode.

The MATLAB function for determining the roots of a polynomial is the **roots** function. The general format for using **roots** is

roots(c)

where **c** is a vector that contains coefficients of the polynomial (coefficients ordered in descending powers of x).

For example, for a polynomial

$$g(x) = x^4 + 3x^3 + 2x + 4 = 0 \qquad (4.24)$$

the coefficients c = [1 3 0 2 4]. The roots are obtained by the statement

    b = roots(c)

The resulting roots are

    b =
        -3.0739
        0.5370 + 1.0064i
        0.5370 - 1.0064i
        -1.0000

If the roots of a polynomial are known and we want to determine the coefficients of the polynomial that corresponds to the roots, we can use the **poly** function. The general form for using **poly** function is

poly(r)

where

r is a vector that contains the results of a polynomial.

poly(r) returns the coefficients of the polynomial whose roots are contained in the vector r.

In the previous example, the roots of the polynomial $x^4 + 3x^3 + 2x + 4 = 0$ were

```
b  =

    -3.0739
    0.5370  +  1.0064i
    0.5370  -  1.0064i
    -1.0000
```

To confirm that the roots will give us the corresponding polynomial, we use the statement

```
g_coeff  =  poly(b')
```

and we obtain

```
g_coeff =
          1.0000   3.000   0.0000   2.0000   4.0000
```

It should be noted that the g_coeff are the same as the coefficients of the polynomial $g(x)$ of Equation (4.24).

The MATLAB function **polyval** is used for polynomial evaluation. The general form of polynomial is

polyval(p, x)

where

p is a vector whose elements are the coefficients of a polynomial in descending powers.

polyval(p, x) is the value of the polynomial evaluated at x.

Evaluate, for example, the polynomial

$$h(x) = 3x^4 + 4x^3 + 5x^2 + 2x + 1 \tag{4.25}$$

At x = 3, we use the statements

```
p = [3 4 5 2 1]
polyval (p, 3)
```

Then we get

```
ans =
        403
```

4.7.2 Zero of a Function (fzero) and Non-zero of a Function (find)

The MATLAB function **fzero** is used to obtain the zero of a function of one variable. The general form of the **fzero** function is

> **fzero('function', x1)**
> **fzero('function', x1, tol)**

where
> **fzero('funct', x1)**   finds the zero of the function funct(x) that is near the point x1.
> **fzero('funct', x1, tol)**   returns zero of the function funct(x) accurate to the relative error of *tol*.

The **find** function determines the indices of the non-zero elements of a vector or matrix. The statement

> C = *find* (*f*)

will return the indices of the vector f that are non-zero. For example, to obtain the points where a change of sign occurs, the statement

```
D = find(product < 0)
```

will show the indices of the locations in product that are negative.

4.7.3 Frequency Response of a Transfer Function (freqs)

The MATLAB function **freqs** is used to obtain the frequency response function H(s). The general form of the function is

$$hs = freqs(num, den, range)$$

where

$$H(s) = \frac{y(s)}{x(s)} = \frac{b_m s^m + b_{m-1} s^{m-1} + \ldots + b_1 s + b_0}{a_n s^n + a_{n-1} s^{n-1} + \ldots + a_1 s + a_0} \qquad (4.26)$$

num $= b_m\, b_{m-1} \ldots b_1\, b_0]$, coefficients of numerator polynomial
den $= [a_n\, a_{n-1} \ldots a_1\, a_0]$, coefficients of denominator polynomials
range is the range of frequencies
hs   is the frequency response (in complex number form)

freqs is an m-file in the MATLAB Signal Processing Toolbox. It is also available in the Student Edition of MATLAB. The polynomial at each frequency point is evaluated. It then divides the numerator response by the denominator response. The **freqs** algorithm is

```
s = sqrt(-1)*w;
h = polyval(b,s)./polyval(a,s);
```

The following example explores the use of the **freqs** function.

Example 4.8   Frequency Response from a Transfer Function

Determine the frequency (magnitude) response of a system whose transfer function is given as

$$H(s) = \frac{4s}{s^2 + 64s + 16} \tag{4.27}$$

Solution

The following MATLAB script can be used to obtain the frequency (magnitude) response.

MATLAB script:

```
%
% Magnitude response of a transfer function
num = [4 0]; % coefficients of numerator polynomial
den = [1 64 16]; % coefficients of denominator polynomial
w = logspace (-4, 5); % range of frequencies
hs = freqs(num, den, w);
f = w/(2*pi); % frequency in Hz.
hs_mag = 20*log10(abs(hs)); % Magnitude in decibels
% Plot the magnitude response
semilogx(f, hs_mag)
title ('Magnitude Response')
xlabel('Frequency, Hz')
ylabel('Magnitude, dB')
```

The frequency (magnitude) response is shown in Figure 4.7.

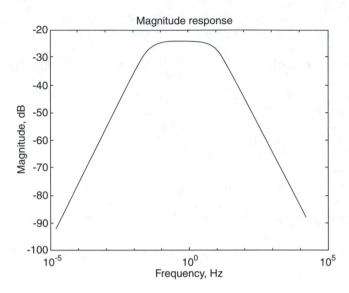

FIGURE 4.7
The magnitude response from a transfer function.

FIGURE 4.8
Block diagram of an amplifier.

The following example uses the MATLAB functions **freqs** and **find** to obtain unity gain frequency of an amplifier.

Example 4.9   Unity Gain Crossover Frequency

The unity gain crossover frequency is the frequency wherein the magnitude of a transfer function is unity. If the transfer function of an amplifier shown in Figure 4.8, is given as

$$H(s) = \frac{2.62 \times 10^{18}}{(s + 400\pi)(s + 8\pi \times 10^5)(s + 1.6\pi \times 10^6)} \tag{4.28}$$

determine the crossover frequency.

Solution

MATLAB script:

```
% Gain crossover frequency
% Transfer function parameters
% poles are
p1 = 400*pi; p2 = 8e5*pi; p3 = 1.6e6*pi;
% determine the coefficients for numerator
% and denominator polynomial
a2 = p1 + p2 + p3;
a1 = p1*p2 + p1*p3 + p2*p3;
a0 = p1*p2*p3;
den = [1 a2 a1 a0]; % coefficients of denominator polynomial
num = [2.62e18];  % coefficients. of numerator polynomial
w = logspace(-1, 7, 5000);  % range of frequencies
hs = freqs(num, den, w);
hs_mag = 20*log10(abs(hs));  % magnitude characteristics
%
f = w/(2*pi);
plot(f,hs_mag)
xlabel('Frequency, Hz')
ylabel('Gain, dB')
title('Frequency Response of an Amplifier')
% gain crossover calculation, unity gain = 0 db gain
lenw = length(w);
lenw1 = lenw - 1;
for i = 1:lenw1
    prod(i) = hs_mag(i)*hs_mag(i+1);
end
fcrit = f(find(prod < 0));
f_cross = fcrit;
fprintf('The crossover frequency is % 9.4e\n', f_cross)
```

The result obtained from MATLAB is: The crossover frequency is 3.2861E+004 Hz.

The plot of the magnitude response is shown in Figure 4.9.

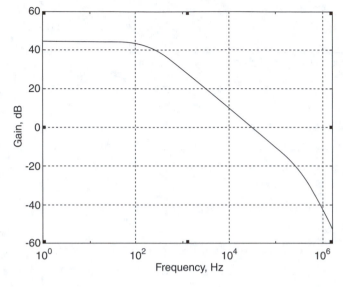

FIGURE 4.9
Magnitude response of an amplifier.

4.8   Save, Load, and Textread Functions

Some input/output commands of MATLAB were discussed in Section 3.9. These were the **break, disp, echo, format, fprintf, input,** and **pause** functions. This section discusses some additional input/output commands, including **save, load,** and **textread**.

4.8.1   Save and Load Functions

The **save** command saves data in MATLAB workspace to disk. The **save** command can store data either in memory-efficient binary format (called a MAT-file) or in an ASCII file. The general form of the **save** command is

> **save filename [List of variables] [options]**

where
> **save** (without filename, list of variables and options), saves all the data in the current workspace to a file named **matlab.mat** in the current directory.
> If a filename is included in the command line, the data will be saved in file **"filename.mat"**.
> If a list of variables is included, only those variables will be saved.

The options for the **save** command are shown in Table 4.5.

TABLE 4.5

Save Command Options

Option	Description
-mat	Save data in MAT-file format (default)
-ascii	Save data using 8-digit ASCII format
-ascii -double	Save data using 16-digit ASCII format
-ascii -double -tab	Saves data using 16-digit ASCII format with tabs
-append	Save data to an existing MAT-file
-v4	Save data in a format that MATLAB Version 4 can open and read

MAT-files are preferable for data that are generated and are going to be used by MATLAB. MAT-files are platform independent. The files can be written and read by any computer that supports MATLAB. In addition, MAT-files preserve all the information about each variable in the workspace, including its name, size, storage space in bytes, and class (structure array, double array, cell array, or character array). Furthermore, MAT-files have every variable stored in full precision.

The ASCII files are preferable if the data are to be exported or imported to programs other than MATLAB. It is recommended that if you save work-space content in ASCII format, *save only one variable at a time*. If more than one variable is saved, MATLAB will create ASCII data that might be difficult to interpret when loaded back in a MATLAB program.

The **load** command will load data from a MAT-file or an ASCII file into the current workspace. The general format of the **load** command is

load filename [options]

where
  load (by itself without filename and options) will load all the data in file matlab.mat into the current workspace.
  load filename will load data from the specified filename.

The options for the **load** command are shown in Table 4.6.

It is strongly recommended that any ASCII data file that will be used with the MATLAB program should contain only numeric information and each row of the file should have the same number of data values. It is also recommended that an ASCII filename include the extension **.dat** so that it may be easier to distinguish between m-files and MAT-files.

TABLE 4.6

Load Command Option

Option	Description
-mat	Load data from MAT-file (default in file extension is mat)
-ascii	Load data from space-separated file

Suppose that a data file stored on disk under a filename **rc_1.dat** contains the data shown in Table 4.7.

TABLE 4.7

Data Stored in File rc_1.dat

Time (s)	Voltage (V)
0.0	0.0
0.5	3.94
1.0	6.32
1.5	7.77
2.0	8.65
2.5	9.18
3.0	9.50
3.5	9.69
4.0	9.82
4.5	9.89
5.0	9.93

The following command

```
load   rc_1.dat  -ascii
```

will load the data into MATLAB and the data will be stored in the matrix rc_1, which has two columns of data. If one gives the command

```
rc_1
```

then one will get:

```
rc_1 =
         0         0
    0.5000    3.9400
    1.0000    6.3200
    1.5000    7.7700
    2.0000    8.6500
    2.5000    9.1800
    3.0000    9.5000
    3.5000    9.6900
    4.0000    9.8200
    4.5000    9.8900
    5.0000    9.9300
```

4.8.2 The Textread Function

The **textread** command can be used to read ASCII files that are formatted into columns of data, where values in each column might be a different type. The general form of the **textread** command is

[a, b, c,...] = textread(filename, format, n)

where

filename is the name of the file to open. The filename should be in quotes; i.e., 'filename'.

format is a string containing a description of the type of data in each column. The format descriptors are similar to those of fprintf (discussed in Section 3.9). The format list should be in quotes. Supported functions include:

  %d — read a signed integer value

  %u — read a integer value

  %f — read a floating point value

  %s — read a whitespace separated string

  %q — read a (possibly double quoted) string

  %c — read characters (including white space) (output is char array)

n is the number of lines to read. If **n** is missing, the command reads to the end of the file.

a,b,c... are the output arguments. The number of output arguments must match the number of columns that are being read.

The **textread** is much more than the **load** command. The **load** command assumes that all the data in the file being load is of single type. The **load** command does not support different data types in different columns. In addition, the **load** command stores all the data in a single array. However, the **textread** command allows each column of data to go into a separate variable.

For example, suppose the file rc_2.dat contains the data shown in Table 4.8. If the first column is time and the second column the voltages across a capacitor, we can use **textread** function to read the data.

TABLE 4.8

Data Stored in File rc_2.dat

Time (s)	Voltage (V)
0.0	50.0
1.0	30.3
2.0	18.4
3.0	11.2
4.0	6.77
5.0	4.10
6.0	2.49
7.0	1.51
8.0	0.916

```
[time,volt_cap]  =  textread('rc_2.dat',  '%f  %f')
time
volt_cap
```

If the above statements are executed, the results are

```
time =

     0
     1
     2
     3
     4
     5
     6
     7
     8
```

```
volt_cap =

     50.0000
     30.3000
     18.4000
     11.2000
      6.7700
      4.1000
      2.4900
      1.5100
      0.9160
```

The following example illustrates the use of the **load** function.

Example 4.10 Statistical Analysis of Data Stored in File

The data shown in Table 4.9 represents two voltages obtained from Monte Carlo analysis of a circuit. V1 and V2 are voltages at two nodes of the circuit. The data is stored in file fproc.dat. (a) Read the data from file and plot V1 as a function of simulation run. (b) Find the mean and standard deviation of V1 and V2.

Solution

Because the **textread** function is not applicable to data in exponential notation, the **load** command can be used to read in the data on file.

TABLE 4.9

Voltages Obtained from Monte Carlo Analysis

Simulation Run	Voltage V1	Voltage V2
1	3.393E+00	8.262E–01
2	3.931E+00	8.483E–01
3	3.761E+00	7.991E–01
4	3.515E+00	8.877E–01
5	3.716E+00	8.922E–01
6	3.243E+00	8.267E–01
7	3.684E+00	7.838E–01
8	3.314E+00	7.687E–01
9	3.778E+00	7.661E–01
10	3.335E+00	9.185E–01
11	3.332E+00	7.991E–01
12	2.993E+00	8.460E–01
13	3.505E+00	7.274E–01
14	3.380E+00	7.873E–01
15	3.584E+00	9.163E–01
16	3.697E+00	7.829E–01
17	3.373E+00	8.119E–01
18	3.106E+00	8.082E–01
19	3.453E+00	7.590E–01
20	3.474E+00	8.647E–01

MATLAB script:

```
% data is stored in fproc.dat
% read data using load command
%
load fproc.dat -ascii
k = fproc(:,1);
v1 = fproc(:,2);
v2 = fproc(:,3);
n = length(k);
% calculate the mean and standard deviation
mean_v1 = mean(v1);   % mean of V1
std_v1 = std(v1);   % standard deviation of v1
mean_v2 = mean(v2);   % mean of V2
std_v2 = std(v2);   % standard deviation of V2
plot(k, v1); % plot of simulation run and V1
xlabel ('Simulation Run')
```

(continued)

```
ylabel('Voltage V1')
title('Plot of Voltage V1')
% Print out results
fprintf('Mean value of V1 is%9.4e volts\n',mean_v1)
fprintf('Standard deviation of V1 is %9.4e volts\n',
std_v1)
fprintf('Mean value of V2 is %9.4e volts\n', mean_v2)
fprintf('Standard deviation of V2 is %9.4evolts\n',
std_v2)
```

The results are

   Mean value of V1 is 3.4784E+000 volts
   Standard deviation of V1 is 2.3638E–001 volts
   Mean value of V2 is 8.2100E–001 volts
   Standard deviation of V2 is 5.3792E–002 volts

Figure 4.10 shows the the voltage V1 as a function of the simulation run.

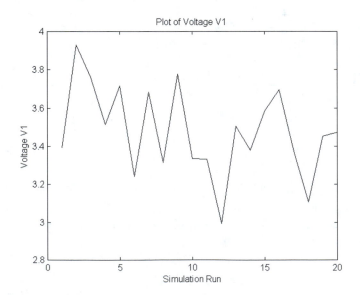

FIGURE 4.10
Voltage V1 as a function of simulation run.

4.9 Interfacing SPICE to MATLAB

As mentioned in Chapter 1, SPICE is the *de facto* standard for circuit simulation. It can perform dc, ac, transient, Fourier, and Monte Carlo analyses. In addition, SPICE has device models incorporated into its package. As well, there is an extensive library of device models available that a SPICE program user can use for simulation and design. As discussed in Section 2.6, PSPICE has an analog behavioral model facility that allows modeling of analog circuit functions using mathematical equations, tables, and transfer functions. The above features of PSPICE are unmatched by other scientific packages, such as MATLAB, MATHCAD, and MATHEMATICA. On the other hand, MATLAB is primarily a tool for matrix computations. It has numerous functions for data processing and analysis. In addition, MATLAB has a rich set of plotting capabilities that are integrated into the MATLAB package. Furthermore, because MATLAB is also a programming environment, a user can extend the MATLAB functional capabilities by writing new modules (m-files).

This book uses the strong features of PSPICE and the powerful functions of MATLAB for electronic circuit analysis. PSPICE can be used to perform dc, ac, transient, Fourier, temperature, and Monte Carlo analyses of electronic circuits with device models and subsystem subcircuits. Then, MATLAB can be used to perform calculation of device parameters, curve fitting, numerical integration, numerical differentiation, statistical analysis, and two- and three-dimensional plots.

PSPICE has the postprocessor package PROBE that can be used for plotting PSPICE results. PROBE also has built-in functions that can be used to do simple signal processing. The valid functions for PROBE expressions are shown in Table 1.4. Compare Table 1.4 of PROBE expressions and Tables 4.1 and 4.2 of MATLAB mathematical and data analysis functions. PSPICE PROBE mathematical expressions do not have the MATLAB functions shown in Table 4.10. It can be seen from Table 4.10 that MATLAB has extensive functions for data analysis, unavailable in PSPICE.

Both PSPICE and MATLAB have functions for performing numerical integration and differentiation. These functions are shown in Table 4.11. MATLAB has several functions for numerical integration. Some of the MATLAB functions allow the user to specify the tolerance limit for stopping the numerical integration (tol). This facility is unavailable in PSPICE.

To exploit the best features of PSPICE and MATLAB, circuit simulation will be done using PSPICE. The PSPICE results, which are written into filename named **filename.out** will be edited using a text editor or a word processor and the data will be saved as **filename.dat**. The data will be read using either MATLAB **textread** or **load** commands. Further processing on the data will be done using MATLAB. The methodology is shown in Figure 4.11.

Subsequent chapters use the methodology described in this section to analyze electronic circuits.

TABLE 4.10

MATLAB Functions Unavailable in PSPICE

Matlab Function	Description
corrcoeff	Obtains correlation coefficients
cov(x)	Determines covariance matrix
cross(x,y)	Finds cross-product of vectors x and y
cumprod(x)	Obtains the cumulative product of columns
cumsum(x)	Obtains the cumulative sum of columns or cumulative sum of elements in a column
hist(x)	Draws the histogram or the bar chart of x using 10 bins
median(x)	Finds the median value of the elements in the vector x
std(x)	Calculates and returns the standard deviation of x
rand(x)	Produces an n-by-n matrix containing normally distributed (Gaussian) random numbers with a mean of 0 and variance of 1
sort(x)	Sorts the rows of a matrix x in ascending order
sum(x)	Calculates and returns the sum of the elements in x
fzero(x)	Finds zero of a function
find(x)	Determines the indices of the non-zero elements of x
polyfit	Determines the polynomial curve fit
fix(x)	Rounds x to the nearest integer toward zero
floor(x)	Rounds x to the nearest integer toward minus infinity
round(x)	Rounds toward nearest integer

TABLE 4.11

Functions for Numerical Integration and Different in PSPICE and MATLAB

Mathematical Operation	PSPICE PROBE Function	MATLAB Function
Numerical integration	s(x)	quad('funct', a, b, tol, trace) quad8('funct', a, b, tol, trace) S2 = trapz(x, y)
Numerical differentiation	d(x)	$f'(x) \cong \dfrac{diff(f)}{diff(x)}$

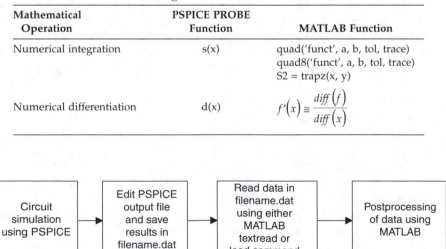

FIGURE 4.11

Flowchart of circuit simulation using PSPICE and postprocessing by MATLAB.

Bibliography

1. Al-Hashimi, Bashir, *The Art of Simulation Using PSPICE, Analog, and Digital,* CRC Press, Boca Raton, FL, 1994.
2. Attia, J.O., *Electronics and Circuit Analysis Using MATLAB,* CRC Press, Boca Raton, FL, 1999.
3. Biran, A. and Breiner, M., *MATLAB for Engineers,* Addison-Wesley, Reading, MA, 1995.
4. Chapman, S. J., *MATLAB Programming for Engineers,* Brook, Cole Thompson Learning, 2000.
5. Derenzo, S.E., *Interfacing: A Laboratory Approach Using the Micrcomputer for Instrumentation, Data Analysis and Control,* Prentice-Hall, Englewood Cliffs, NJ, 1990.
6. Etter, D.M., *Engineering Problem Solving with MATLAB,* 2nd edition, Prentice-Hall, Upper Saddle River, NJ, 1997.
7. Etter, D.M., Kuncicky, D.C., and Hull, D., *Introduction to MATLAB 6,* Prentice-Hall, Upper Saddle River, NJ, 2002.
8. Fenical, L.H., *PSPICE: A Tutorial,* Prentice-Hall, Englewood Cliffs, NJ, 1992.
9. Gottling, J.G., *Matrix Analysis of Circuits Using MATLAB,* Prentice-Hall, Englewood Cliffs, NJ, 1995.
10. Lamey, Robert, *The Illustrated Guide to PSPICE,* Delmar Publishers, Albany, NY, 1995.
11. Monssen, Franz, *PSPICE with Circuit Analysis,* MacMillan Publishing, New York, 1992.
12. Rashid, M.H., *Microelectronic Circuits,* PWS Publishing, Boston, MA, 1999.
13. Sedra, A.S. and Smith, K.C., *Microelectronic Circuits,* 4th edition, Oxford University Press, New York, 1998.
14. Sigmor, K., *MATLAB Primer,* 4th edition, CRC Press, Boca Raton, FL, 1998.
15. Using MATLAB, The Language of Technical Computing, Computation, Visualization, and Programming, Version 6, MathWorks, Inc., 2000.

Problems

4.1 A triangular wave, symmetric about t = 0 has a Fourier series expansion

$$x(t) = \frac{8A}{\pi^2}\left(\cos \omega t + \frac{1}{9}\cos 3\omega t + \frac{1}{25}\cos 5\omega t + \frac{1}{49}\cos 7\omega t + \ldots\right)$$

where A is peak value and ω is the angular frequency. If $A = 2$ and $\omega = 400\pi$ rad/s, write a MATLAB program to synthesize the triangular wave using the first 4 terms.

4.2 A square wave with a peak to peak value of 4 V and an average value of 0 V can be expressed as

$$x(t) = \frac{8}{\pi}\left(\cos\omega_0 t - \frac{1}{3}\cos 3\omega_0 t + \frac{1}{5}\cos 5\omega_0 t - \frac{1}{7}\cos 7\omega_0 t + \dots \right)$$

where $\omega = 2\pi f_0$ and $f_0 = 100$ Hz, write a MATLAB program to synthesize the square wave using the first 4 terms.

4.3 The gains β of fifteen 2N2907 transistors, measured in the laboratory, are 170, 200, 160, 165, 175, 155, 210, 190, 180, 165, 195, 200, 195, 205, 190. (a) What is the mean value of β? (b) What are the minimum and maximum value of β? (c) What is the standard deviation of β?

4.4 The Zener voltage regulator circuit shown in Figure P4.4 has RS = 250 Ω and RL = 350 Ω. The diode D1 is D1N750. The corresponding input and output voltages are shown in Table P4.4. (a) Find the minimum, maximum, mean, and standard deviation of the input and output voltages. (b) What is the correlation coefficient between input and output voltages? (c) What is the least mean square fit (polynomial fit)?

TABLE P4.4

Input and Output Voltages of Zener
Voltage Regulator

Input Voltage VS (V)	Output Voltage VOUT (V)
8.0	4.601
9.0	4.658
10.0	4.676
11.0	4.686
12.0	4.694
13.0	4.699
14.0	4.704
15.0	4.708
16.0	4.712
17.0	4.715
18.0	4.717

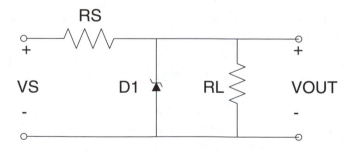

FIGURE P4.4
Zener diode shunt voltage regulator circuit.

4.5 The transfer function of a bandpass filter is

$$H(s) = \frac{s\,R/L}{s^2 + s\,R/L + 1/LC}$$

If L = 5 mH, C = 20 μF, and R = 15KΩ, (a) use MATLAB to plot the magnitude response, and (b) find the frequency at which the maximum value of the magnitude response occurs.

4.6 The differentiator circuit shown in Figure P4.6(a) has R = 10 KΩ and C = 10 μF. The input voltage is the sawtooth waveform shown in Figure P4.6(b). Use MATLAB to plot the output waveform. Calculate the mean value and rms value of the output voltage $v_0(t)$.

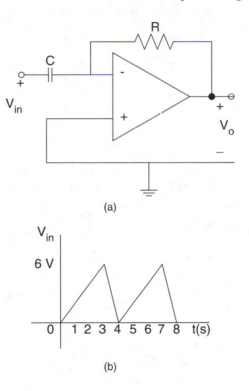

(a)

(b)

FIGURE P4.6
(a) Integrator circuit and (b) sawtooth waveform.

4.7 A static random access memory (SRAM) operating in a radiation environment can experience a change in logic state. For the currents at various instants of time at a node of an SRAM shown in Table P4.7, calculate the charge deposited on the node.

TABLE P4.7

Current at SRAM Node

Time (s)	Current (A)
0.000E+00	0.000E+00
2.000E–08	0.000E+00
4.000E–08	4.000E–12
6.000E–08	3.672E–02
8.000E–08	3.970E–02
1.000E–07	3.998E–02
1.200E–07	4.000E–02
1.400E–07	4.000E–02
1.600E–07	3.281E–02
1.800E–07	2.200E–02
2.000E–07	1.475E–02
2.200E–07	9.887E–03
2.400E–07	6.628E–03
2.600E–07	4.442E–03
2.800E–07	2.978E–03
3.000E–07	1.996E–03
3.200E–07	1.338E–03
3.400E–07	8.969E–04
3.600E–07	6.013E–04
3.800E–07	4.030E–04
4.000E–07	2.695E–04

4.8 In a platinum resistance thermometer with the American Alloy, the resistance R is related to the temperature T by the expression

$$R = a + bT + cT^2$$

where R is the resistance in ohms, T is temperature in °C. Using the data in Table P4.8, determine the coefficients a, b, and c.

4.9 To the first order in $1/T$, the relationship between resistance R and temperature T of a thermistor is given by

$$R(T) = R(T_0) \exp\left(\beta\left(\frac{1}{T} - \frac{1}{T_0} \right) \right)$$

where

T is in degrees Kelvin

T_0 is the reference temperature, in degrees Kelvin

$R(T_0)$ is the resistance at the reference temperature

β is the temperature coefficient

TABLE P4.8

Temperature vs. Resistance
of a Thermometer

T (°C)	R (Ω)
−100	60.30
−80	68.34
−60	76.32
−40	84.26
−20	92.16
0	100
20	107.79
40	115.54
60	123.24
80	130.89
100	138.50

If $R(T_0)$ is 10,000 Ω at the reference temperature, use Table P4.9 to obtain the constants β and T_0.

TABLE P4.9

Temperature vs. Resistance
of a Thermistor

T (°K)	R (Ω)
240	1.7088E+005
260	0.5565E+005
280	0.2128E+005
300	0.0925E+005
320	0.0446E+005
340	0.0234E+005
360	0.0132E+005
380	0.0079E+005
400	0.0050E+005
420	0.0033E+005
440	0.0023E+005

4.10 An amplifier has the following voltage transfer function.

$$A(s) = \frac{150s}{\left(1 + \dfrac{s}{10^3}\right)\left(1 + \dfrac{s}{5 \times 10^4}\right)}$$

(a) Plot the magnitude vs. the frequency. (b) Find the gain crossover frequency, fgc, where the magnitude becomes unity.

4.11 The voltage transfer function of an uncompensated op amp is given as:

$$A(s) = \frac{250}{\left(1 + \frac{s}{200\pi}\right)} \frac{100}{\left(1 + \frac{s}{80\pi}\right)} \frac{0.8}{\left(1 + \frac{s}{2.5\pi \times 10^7}\right)}$$

(a) Draw the Bode plot for the magnitude vs. frequency. (b) Find the gain-crossover frequency (fgc) where the magnitude of A(s) becomes unity.

4.12 The voltage between two nodes of circuit for different supply voltages is shown in Table P4.12. Plot V1 vs. V2. Determine the line of best fit between V1 and V2.

TABLE P4.12

Voltages at Two Nodes

Voltage V1 (V)	Voltage V2 (V)
5.638	5.294
5.875	5.644
6.111	5.835
6.348	6.165
6.584	6.374
6.820	6.684
7.055	6.843
7.290	7.162
7.525	7.460
7.759	7.627
7.990	7.972
8.216	8.170
8.345	8.362

Section III

Applications of PSPICE and MATLAB

5

Diode Circuits

This chapter discusses diode circuits. The chapter begins with a discussion of diode characteristics, followed by diode rectification. Peak detector and limiter circuits that use diodes are discussed, as are Zener diodes and Zener voltage regulator circuits. Most of the examples in the chapter are done using both PSPICE and MATLAB.

5.1   Diode

In the forward-biased and reversed-biased regions, the current i_D and the voltage, v_D of a semiconductor diode are related by the diode equation

$$i_D = I_S \left[e^{(v_D/nV_T)} - 1 \right] \tag{5.1}$$

where
   I_S is reverse saturation current
   n is an empirical constant between 1 and 2
   V_T is thermal voltage, given by

$$V_T = \frac{kT}{q} \tag{5.2}$$

   k is the Boltzman constant $= 1.38 \times 10^{-23}$ J/°K
   q is the electronic charge $= 1.6 \times 10^{-19}$ coulombs
   T is the absolute temperature in °K

In the forward-biased region, the voltage across the diode is positive. Assuming that the voltage across the diode is greater than 0.5 V, Equation (5.1) simplifies to

$$i_D = I_s e^{v_D/nV_T} \tag{5.3}$$

From Equation (5.3), we get

$$\ln(i_D) = \frac{v_D}{nV_T} + \ln(I_s) \tag{5.4}$$

For a particular operating point of the diode ($i_D = I_D$ and $v_D = V_D$), we can obtain the dynamic resistance of the diode r_d at a specified operating point as

$$r_d = \frac{di_D}{dv_D}\bigg|_{v_D=V_D} = \frac{I_s e^{(V_D/nV_T)}}{nV_T} \tag{5.5}$$

Equation (5.4) can be used to obtain the diode constants n and I_s, given a data that consists of the corresponding values of voltage and current. From Equation (5.4), a curve of v_D versus $\ln(i_D)$ will have a slope given by $1/nV_T$ and a y-intercept of $\ln(I_s)$. The following example illustrates how to find n and I_s from experimental data. The example uses the MATLAB function **polyfit**, which was discussed in Chapter 4.

Example 5.1 Determination of Diode Parameters from Experimental Data

A forward-biased semiconductor diode has the corresponding voltages and currents shown in Table 5.1. Determine the reverse saturation current I_s and the diode constant n. Plot the line of best fit.

TABLE 5.1

Current vs. Voltage of a Forward-Biased Diode

Forward-Biased Voltage v_D (V)	Forward-Biased Current i_D (A)
0.2	6.37E–9
0.3	7.75E–8
0.4	6.79E–7
0.5	3.97E–6
0.6	5.59E–5
0.7	3.63E–4

Solution

MATLAB script:

```
% Diode parameters
vt = 25.67e-3;
vd = [0.2 0.3 0.4 0.5 0.6 0.7];
id = [6.37e-9 7.75e-8 6.79e-7 3.97e-6 5.59e-5 3.63e-4];
%
lnid = log(id);   % Natural log of current
% Determine coefficients
pfit = polyfit (vd, lnid, 1); % curve fitting
% Linear equation is y = mx + b
b = pfit(2);
m = pfit(1);
ifit = m*vd + b;
% Calculate Is and n
Is = exp(b)
n = 1/(m*vt)
% Plot v versys ln(i) and best fit linear model
plot(vd, ifit, 'b', vd, lnid, 'ob')
xlabel ('Voltage, V')
ylabel ('ln(i)')
title ('Best Fit Linear Model')
```

The results obtained from MATLAB are

Is = 9.5559E–011

n = 1.7879

Figure 5.1 shows the best-fit linear model used to determine the reverse saturation current I_S and the diode parameter n.

In the diode Equation (5.1), the thermal voltage V_T and the reverse saturation current I_S are temperature dependent. The thermal voltage is directly proportional to temperature. This is shown in Equation (5.2). The reverse saturation current increases approximately 7.2%/°C for both silicon and germanium diodes. The expression for the reverse saturation current as a function of temperature is

$$I_S(T_2) = I_S(T_1)e^{[k_s(T_2-T_1)]} \tag{5.6}$$

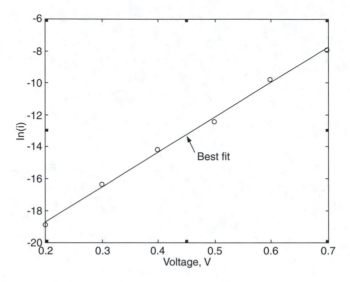

FIGURE 5.1
Best-fit linear model.

where
  $k_S = 0.072/°C$
  T_1 and T_2 are two different temperatures

The following example shows the effect of the temperature on the output voltage of a diode circuit.

Example 5.2   Temperature Effects on a Diode

For the circuit shown in Figure 5.2, VS = 5 V, R1 = 50 KΩ, R2 = 20 KΩ, and R3 = 50 KΩ. If the diode D1 is D1N4009, plot temperature vs. output voltage. Find the equation of best fit between the voltage and the temperature.

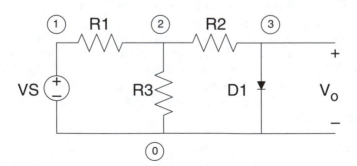

FIGURE 5.2
Diode circuit.

Solution

PSPICE is used to obtain the voltage at various temperatures. The SPICE command **.TEMP** is used to obtain the temperature effects. MATLAB is used to plot the relationship between temperature and diode voltage.

PSPICE program:

```
DIODE CIRCUIT
VS            1       0       DC      5V
R1            1       2       50E3
R2            2       0       20E3
R3            2       3       50E3
D1            3       0       D1N4009
.MODEL D1N4009 D(IS=0.1P RS=4 CJO=2P TT=3N BV=60 IBV=0.1P)
.STEP                TEMP    0       100     10
.DC          VS              5       5       1
.PRINT DC    V(3)
.END
```

The PSPICE results showing the temperature vs. voltage across the diode can be found in file ex5_2ps.dat. The results are shown in Table 5.2.

TABLE 5.2

Diode Voltage vs. Temperature

Temperature (°C)	Diode Voltage (V)
0	0.5476
10	0.5250
20	0.5023
30	0.4796
40	0.4568
50	0.4339
60	0.4110
70	0.3881
80	0.3651
90	0.3421
100	0.3190

MATLAB script:

```
% Processing of PSPICE data using MATLAB
% Read data using textread command
%
```

(continued)

```
[temp, vdiode]=textread ('ex5_2ps.dat', '%d %f');
vfit=polyfit(temp,vdiode, 1);
% Linear equation is y=mx + b
b=vfit(2)
m=vfit(1)
vfit=m*temp + b;
plot(temp, vfit, 'b', temp, vdiode, 'ob');% plot
temperature vs. diode voltage
xlabel ('Temperature in °C')
ylabel ('Voltage, V')
title ('Temperature versus Diode Voltage')
```

From the above MATLAB program for plotting voltage vs. temperature,

   b =

        0.5480

   m =

        −0.0023

The equation of best fit between temperature and voltage is

$$v_O(T) = -0.0023T + 0.548 \text{ V}$$

The plot is shown in Figure 5.3.

5.2   Rectification

A half-wave rectifier circuit is shown in Figure 5.4.

It consists of an alternating current (ac) source, a diode, and a resistor. Assuming that the diode is ideal, the diode conducts when the source is positive, making

$$v_0 = v_S \text{ when } v_S > 0 \tag{5.7}$$

When the source voltage is negative, the diode is cut-off, and the output voltage is

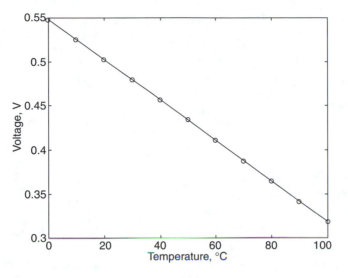

FIGURE 5.3
Diode voltage vs. temperature.

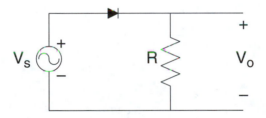

FIGURE 5.4
Half-wave rectifier circuit.

$$v_0 = 0 \quad \text{when} \quad v_S < 0 \tag{5.8}$$

A battery charging circuit is a slight modification of the half-wave rectifier circuit. The battery charging circuit, explored in the following example consists of an alternating current source connected to a battery through a resistor and a diode.

Example 5.3   Battery Charging Circuit

In the battery charging circuit shown in Figure 5.5, VB = 12 V and R = 50 Ω. The source voltage is $v_s(t) = 16 \sin(120 \, \pi t)$ V. If the diode is D1N4448, plot the current flowing through the diode and determine the peak current and average current flowing into the battery. What is the total charge supplied to the battery?

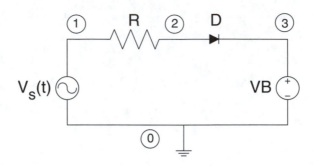

FIGURE 5.5
Battery charging circuit.

Solution

PSPICE is used to obtain the output current as a function of time. Three cycles of the input signal are used in the simulations. The resulting data are analyzed using MATLAB.

PSPICE program:

```
*  BATTERY CHARGING CIRCUIT
VS         1        0        SIN(0 16 60)
R                   1        2        50
D                            2        3        D1N4448
VB         3        0        DC      12V
.MODEL D1N4448 D(IS=0.1P RS=2
4 CJO=2P TT=12N BV=100 IBV=0.1P)
.TRAN    0.5MS    50MS    0        0.5MS
.PRINT              TRAN I(R)
.PROBE
.END
```

The PSPICE results showing the current flowing through the diode can be found in file ex5_3ps.dat. The partial results are in shown in Table 5.3.

The MATLAB program for the analysis of the PSPICE results follows.

TABLE 5.3

Current Flowing Through a Diode

Time (s)	Current (A)
0.000E+00	−1.210E−11
1.000E−03	4.201E−09
2.000E−03	6.160E−09
3.000E−03	3.385E−02
4.000E−03	6.148E−02
5.000E−03	4.723E−02
6.000E−03	2.060E−03
7.000E−03	−4.583E−09
8.000E−03	−3.606E−09
9.000E−03	−2.839E−09
1.000E−02	−2.070E−09
1.100E−02	−1.268E−09
1.200E−02	−4.366E−10
1.300E−02	3.940E−10
1.400E−02	1.231E−09
1.500E−02	2.041E−09
1.600E−02	2.820E−09
1.700E−02	3.601E−09

MATLAB script:

```
% Battery charging circuit
%
% Read PSPICE results using load function
load 'ex5_3ps.dat' -ascii;
time = ex5_3ps(:,1);
idiode = ex5_3ps(:,2);
plot(time, idiode), % plot of diode current
xlabel('Time, s')
ylabel('Diode Current, A')
title('Diode Current as a Function of Time')
i_peak = max(idiode);  % peak current
i_ave = mean(idiode);  % average current
% Function trapz is used to integrate the current
charge = trapz(time, idiode);
% Print out the results
fprintf ('Peak Current is % 10.5e A\n', i_peak)
fprintf('Average current is % 10.5e A\n', i_ave)
fprintf('Total Charge is %10.5e C\n', charge)
```

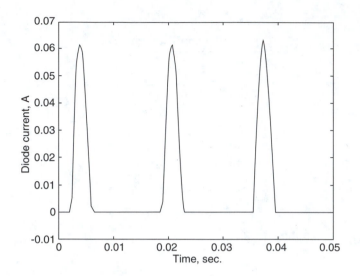

FIGURE 5.6
Diode current.

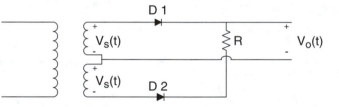

FIGURE 5.7
Full-wave rectifier with center-tapped transformer.

The results from MATLAB are

Peak Current is 6.32700E–002 A

Average current is 8.66958E–003 A

Total Charge is 4.37814E–004 C

The plot of the current vs. time is shown in Figure 5.6.

A full-wave rectifier that uses a center-tapped transformer is shown in Figure 5.7.

When $v_s(t)$ is positive, the diode D1 conducts but diode D2 is off, and the output voltage is

$$v_0(t) = v_s(t) - v_D \qquad (5.9)$$

where v_D is a voltage drop across a diode.

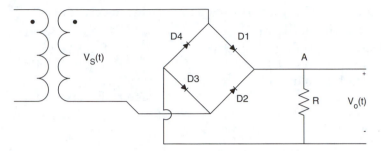

FIGURE 5.8
Bridge rectifier.

When $v_s(t)$ is negative, diode D1 is cut-off but diode D2 conducts. The current flowing through the load R enters the latter through node A. The output voltage is

$$v_s(t) = |v_s(t)| - v_D \qquad (5.10)$$

A full-wave rectifier that does not require a center-tapped transformer is the bridge rectifier. It is shown in Figure 5.8.

When $v_s(t)$ is negative, diodes D2 and D4 conduct but diodes D1 and D3 do not conduct. The current entering load resistance R enters it through node A. The output voltage is

$$v_0(t) = |v_s(t)| - 2v_D \qquad (5.11)$$

When $v_s(t)$ is positive, diodes D1 and D3 conduct but diodes D2 and D4 do not conduct. The current entering the load resistance R enters through node A. The output voltage is given by Equation (5.11).

Connecting a capacitor across the load can smooth the output voltage of a full-wave rectifier. The resulting circuit is shown in Figure 5.9. The following example explores some characteristics of the smoothing circuit.

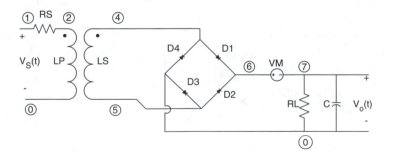

FIGURE 5.9
Full-wave rectifier with capacitor smoothing filter.

Example 5.4　Characteristics of a Bridge Rectifier with Smoothing Filter

For Figure 5.9, $v_s(t) = 120\sqrt{2}\sin(2\pi 60t)$, $C = 100$ μF, RS = 1 Ω, LP = 2 H, LS = 22 mH, and diodes D1, D2, D3, and D4 are D1N4150. In addition, the coefficient of coupling of the transformer is 0.999. Determine the average output voltage and rms value of ripple voltage as RL takes the following values: 10, 30, 50, 70, 90 KΩ. Plot the average output voltage and rms value of output voltage as a function of RL.

Solution

The PSPICE command **.STEP** is used to vary the element values of resistance RL. The diode current and the output voltage are obtained from PSPICE simulation. Further analysis of the voltage and current is done using MATLAB.

PSPICE program:

```
BRIDGE RECTIFIER
*  SINUSOIDAL TRANSIENT INPUT
VS        1        0        SIN(0 169V 60HZ)
RS        1        2        1
*  TRANSFORMER SECTION
LP        2        0        2H
LS        4        5        22MH
KFMR      LP       LS       0.999
*RECTIFIER DIODES
D1        4        6        D1N4150
D2        5        6               D1N4150
D3        0        5               D1N4150
D4        0        4               D1N4150
.MODEL D1N4150 D(IS=10E-15 RS=1.0 CJO=1.3P TT=12N BV=70
IBV=0.1P)
*DIODE CURRENT MONITOR
VM        6        7        DC       0
RL        7        0        RMOD     1
C         7        0        100E-6
.MODEL RMOD RES(R=1)
*  ANALYSIS REQUESTS
.TRAN     0.2MS    100MS
.STEP     RES      RMOD(R)  10K  90K  20K
```

(continued)

```
.PRINT   TRAN    V(7)
.PROBE  V(7)
.END
```

Partial PSPICE results showing the output voltage for RL = 10 KΩ and RL = 90 KΩ are shown in Tables 5.4 and 5.5, respectively.

TABLE 5.4

Output Voltage for a Load
Resistance of 10 KΩ

Time (s)	Output Voltage for RL = 10 KΩ (V)
5.000E–03	1.620E+01
1.000E–02	1.612E+01
1.500E–02	1.604E+01
2.000E–02	1.596E+01
2.500E–02	1.611E+01
3.000E–02	1.622E+01
3.500E–02	1.614E+01
4.000E–02	1.616E+01
4.500E–02	1.608E+01
5.000E–02	1.616E+01
5.500E–02	1.617E+01
6.000E–02	1.609E+01
6.500E–02	1.620E+01
7.000E–02	1.611E+01

TABLE 5.5

Output Voltage for a Load
Resistance of 90 KΩ

Time (s)	Output Voltage for RL = 90 KΩ (V)
5.000E–03	1.621E+01
1.000E–02	1.620E+01
1.500E–02	1.619E+01
2.000E–02	1.619E+01
2.500E–02	1.618E+01
3.000E–02	1.617E+01
3.500E–02	1.616E+01
4.000E–02	1.615E+01
4.500E–02	1.615E+01
5.000E–02	1.614E+01
5.500E–02	1.621E+01
6.000E–02	1.620E+01
6.500E–02	1.620E+01
7.000E–02	1.619E+01

The complete results for RL = 10 , 30, 50, 70, and 90 KΩ can be found in files ex5_4aps.dat, ex5_4bps.dat, ex5_4cps.dat, ex5_4dps.dat, and ex5_4eps.dat, respectively. In the data analysis, the output voltage stabilizes after 5 ms. The output voltages for times less than 5 ms have been deleted in calculating the ripple voltage parameters because the output voltage stabilizes after 5 ms. The MATLAB program for the analysis of the PSPICE results follows.

MATLAB script:

```
% Read PSPICE results using textread
% Load resistors
load 'ex5_4aps.dat' -ascii;
load 'ex5_4bps.dat' -ascii;
load 'ex5_4cps.dat' -ascii;
load 'ex5_4dps.dat' -ascii;
load 'ex5_4eps.dat' -ascii;
t2 = ex5_4aps(:,1);
v10k = ex5_4aps(:,2);
v30k = ex5_4bps(:,2);
v50k = ex5_4cps(:,2);
v70k = ex5_4dps(:,2);
v90k = ex5_4eps(:,2);

rl(1) = 10e3; rl(2) = 30e3; rl(3) = 50e3; rl(4) = 70e3;
rl(5) = 90e3;
%
% Average DC Voltage calculation
v_ave(1) = mean (v10k);
v_ave(2) = mean (v30k);
v_ave(3) = mean (v50k);
v_ave(4) = mean (v70k);
v_ave (5) = mean(v90k);
%
% RMS voltage calculation
n = length(v10k)
for i=1: n
   V1diff(i) = (v10k(i) - v_ave(1))^2; % ripple voltage squared
   V2diff(i) = (v30k(i) - v_ave(2))^2; % ripple voltage squared
```

(continued)

```
    V3diff(i) = (v50k(i) - v_ave(3))^2; % ripple voltage squared
    V4diff(i) = (v70k(i) - v_ave(4))^2; % ripple voltage squared
    V5diff(i) = (v90k(i) - v_ave(5))^2; % ripple voltage squared
end
%
% Numerical Integration
Vint1 = trapz(t2, V1diff);
Vint2 = trapz(t2, V2diff);
Vint3 = trapz(t2, V3diff);
Vint4 = trapz(t2, V4diff);
Vint5 = trapz(t2, V5diff);
%
tup = t2(n);   % Upper Limit of integration
v_rms(1) = sqrt(Vint1/tup);
v_rms(2) = sqrt(Vint2/tup);
v_rms(3) = sqrt(Vint3/tup);
v_rms(4) = sqrt(Vint4/tup);
v_rms(5) = sqrt(Vint5/tup);
%
% plot the average voltage
subplot(211)
plot(rl,v_ave)
ylabel('Average Voltage, V')
title('Average Voltage as a Function of Load Resistance')
% Plot rms voltage vs. RL
subplot(212)
plot(rl, v_rms)
title('Rms Voltage as a Function of Load Resistance')
ylabel('RMS Value')
xlabel('Load Resistance')
```

Figure 5.10 shows the MATLAB plots for the RMS voltage and average diode current.

Figure 5.10 shows that, in general, as the load resistance increases, the average voltage of the output increases and the rms value of the output ripple voltage decreases.

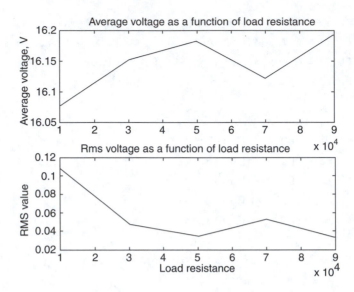

FIGURE 5.10
Average voltage (upper) and RMS voltage (lower) as a function of load resistance.

5.3　Zener Diode Voltage Regulator

A Zener diode is a pn-junction with a controlled breakdown voltage. The current-voltage characteristics of a Zener diode are shown in Figure 5.11.

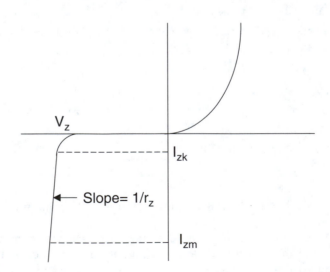

FIGURE 5.11
I–V characteristics of a Zener diode.

I_{zk} is the minimum current needed for the Zener diode to break down. I_{zm} is the maximum current that can flow through the Zener diode without being destroyed. It is obtained by

$$I_{zm} = \frac{P_z}{V_z} \tag{5.12}$$

where
  P_z is the Zener power dissipation
  V_z is the Zener breakdown voltage.

The incremental resistance of the Zener diode at an operating point is specified by

$$r_z = \frac{\Delta V_z}{\Delta I_z} \tag{5.13}$$

The following example explores the incremental resistance of a Zener diode.

Example 5.5 Zener Diode Resistance

The circuit shown in Figure 5.12 can be used to obtain the dynamic resistance of a Zener diode. Assuming that D1 is D1N4742, RS = 2 Ω, RL = 100 Ω, and 12.2 ≤ VS ≤ 13.2, determine the dynamic resistance as a function of the diode voltage.

Solution

PSPICE can be used to obtain Zener current and voltage as the input voltage is varied. MATLAB is used to obtain the dynamic resistance at various operating points of the Zener diode.

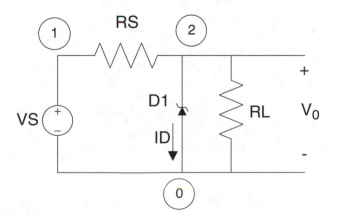

FIGURE 5.12
Zener diode circuit.

PSPICE program:

```
DYNAMIC RESISTANCE OF ZENER DIODE
.OPTIONS RELTOL=1.0E-08
.OPTIONS NUMDGT=6
VS   1   0   DC 12V
RS   1   2   2
D1   0   2   D1N4742
RL   2   0   100
.MODEL D1N4742 D(IS=0.05UA RS=9 BV=12 IBV=5UA)
.DC VS 12.2 13.2 0.05
.PRINT DC V(0,2) I(D1)
.END
```

Partial PSPICE results are shown in Table 5.6. The complete results can be found in file ex5_5ps.dat.

TABLE 5.6

Voltage vs. Current
of Zener Diode

Voltage (V)	Current (A)
−1.19608E+01	−1.14718E−06
−1.20098E+01	−7.33186E−06
−1.20587E+01	−4.76843E−05
−1.21073E+01	−2.86514E−04
−1.21544E+01	−1.26124E−03
−1.21992E+01	−3.39609E−03
−1.22424E+01	−6.38033E−03
−1.22846E+01	−9.83217E−03
−1.23264E+01	−1.35480E−02
−1.24090E+01	−2.14124E−02
−1.24501E+01	−2.54748E−02
−1.24910E+01	−2.95934E−02
−1.25319E+01	−3.37552E−02
−1.25726E+01	−3.79510E−02
−1.26134E+01	−4.21743E−02
−1.26541E+01	−4.64202E−02
−1.26947E+01	−5.06850E−02
−1.27354E+01	−5.49660E−02
−1.27760E+01	−5.92607E−02
−1.28165E+01	−6.35675E−02

MATLAB script:

```
% Dynamic resistance of Zener diode
%
% Read the PSPICE results
load 'ex5_5ps.dat' -ascii;
vd = ex5_5ps(:,2);
id = ex5_5ps(:,3);
n = length(vd); % number of data points
m = n-2;  %number of dynamic resistances to calculate
for i = 1: m
    vpt(i) = vd(i + 1);
    rd(i) = -(vd(i+2) - vd(i))/(id(i+2)-id(i));
end
% Plot the dynamic resistance
plot (vpt, rd,'ob',vpt,rd)
title('Dynamic Resistance of a Zener Diode')
xlabel('Voltage, V')
ylabel('Dynamic Resistance, Ohms')
```

The MATLAB plot is shown in Figure 5.13.

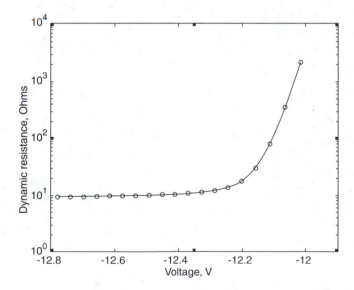

FIGURE 5.13
Dynamic resistance of a Zener diode.

From Figure 5.13, it seems the dynamic resistance reduces at the breakdown region of the Zener.

One of the applications of a Zener diode is its use in the design of voltage reference circuits. A Zener diode shunt voltage regulator is shown in Figure 5.12. The circuit is used to provide an output voltage, V_0, which is nearly constant. When the source voltage is greater than the Zener breakdown voltage, the Zener will break down and the output voltage will be equal to the Zener breakdown voltage.

The output voltage will change slightly with the variation of the load resistance. This is termed output voltage regulation. The following example illustrates the change of output voltage as the load resistance is changed.

Example 5.6   Voltage Regulation of Zener Diode

In Figure 5.12, VS = 18V and RS = 400 Ω. Find the output voltage as RL varies from 1 KΩ to 51 KΩ. The Zener diode is D1N4742.

Solution

PSPICE is used to obtain the output voltage for the various values of load resistance. MATLAB is used to obtain the plot of output voltage as a function of the load resistance.

PSPICE program:

```
VOLTAGE REGULATOR CIRCUIT
.OPTIONS RELTOL=1.0E-08
.OPTIONS NUMDGT=5
VS   1   0   DC 18V
RS 1 2 400
DZENER 0 2 D1N4742
.MODEL D1N4742 D(IS=0.05UA RS=9 BV=12 IBV=5UA)
RL   2   0  ·RMOD 1
.MODEL RMOD RES(R=1)
.STEP RES RMOD(R)  1K 51K 5K
*ANALYSIS TO BE DONE
.DC VS 18 18 1
.PRINT DC V(2)
.END
```

PSPICE results are shown in Table 5.7.

The PSPICE results are also stored in file ex5_6ps.dat. The MATLAB script for plotting the PSPICE results follows.

TABLE 5.7

Output Resistance vs. Output Voltage

Output Resistance RL (Ω)	Output Voltage (V)
1.0000E+03	1.2181E+01
6.0000E+03	1.2311E+01
11.0000E+03	1.2321E+01
16.0000E+03	1.2325E+01
21.0000E+03	1.2327E+01
26.0000E+03	1.2328E+01
31.0000E+03	1.2329E+01
36.0000E+03	1.2329E+01
41.0000E+03	1.2330E+01
46.0000E+03	1.2330E+01
51.0000E+03	1.2331E+01

MATLAB script:

```
% Voltage Regulation
% Plot of Output voltage versus load resistance
% Input the PSPICE results
load 'ex5_6ps.dat' -ascii;
rl = ex5_6ps(:,1);
v = ex5_6ps(:,2);
% Plot rl versus v
plot(rl, v, 'b', rl, v, 'ob')
xlabel('Load Resistance, Ohms')
ylabel('Output Voltage, V')
title('Output Voltage as a Function of Load Resistance')
```

The voltage regulation plot is shown in Figure 5.14. As the load resistance increases, the output voltage becomes almost constant.

Example 5.7 3-D Plot of Voltage Regulation

In Figure 5.13, the load resistance is varied from 200 Ω to 2000 Ω and the source voltage varies from 4 V to 24 V, while RS is kept constant at 150 Ω. Obtain the output voltage with respect to both the load resistance and the source voltage. Assume that the Zener diode is D1N754.

Solution

PSPICE is used to obtain the output voltage as both the source voltage and load resistance are changed.

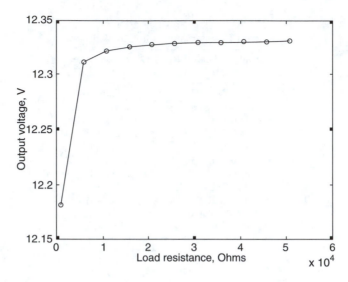

FIGURE 5.14
Load resistance vs. output voltage.

PSPICE program:

```
VOLTAGE REGULATOR CIRCUIT
VS   1   0   DC 18V
RS   1   2   150
DZENER 0 2 D1N754
.MODEL D1N754 D(IS=880.5E-18 N=1 RS=0.25 IKF=0 XTI=3
EG=1.11
+ CJO=175P M=0.5516 VJ=0.75 FC=0.5 ISR=1.859N NR=2
BV=6.863
+ IBV=0.2723 TT=1.443M)
RL   2   0   RMOD 1
.MODEL RMOD RES(R=1)
.STEP RES RMOD(R) 0.2E3 2.0E3 0.2E3
* ANALYSIS TO BE DONE
.DC VS 4   24 1
.PRINT DC V(2)
.END
```

Partial results from PSPICE simulation are shown in Table 5.8. The complete results can be found in the file ex5_7ps.dat.

TABLE 5.8

Output Voltage as a Function of Load Resistance
and Input Voltage

Source Voltage (V)	Load Resistance (Ω)	Output Voltage (V)
4.000E+00	200	2.286E+00
8.000E+00	200	4.571E+00
1.200E+01	200	6.729E+00
1.600E+01	200	6.810E+00
2.000E+01	200	6.834E+00
2.400E+01	200	6.851E+00
4.000E+00	400	2.909E+00
8.000E+00	400	5.818E+00
1.200E+01	400	6.797E+00
1.600E+01	400	6.827E+00
2.000E+01	400	6.846E+00
2.400E+01	400	6.861E+00
4.000E+00	600	3.200E+00
8.000E+00	600	6.400E+00
1.200E+01	600	6.805E+00
1.600E+01	600	6.831E+00
2.000E+01	600	6.849E+00
2.400E+01	600	6.864E+00

MATLAB is used to do the 3-D plot.

MATLAB script:

```
% 3-D plot of output voltage as a function
% of load resistance and input voltage
%
load 'ex5_7ps.dat' -ascii;
vs = ex5_7ps(:,1);
rl = ex5_7ps(:,2);
vo = ex5_7ps(:,3);
% Do 3-D plot
plot3(vs, rl, vo,'r');
% axis square
grid on
title ('Output Voltage as a Function of Load and Source
Voltage')
xlabel('Input Voltage, V')
ylabel('Load Resistance, Ohms')
zlabel('Voltage across zener diode,V')
```

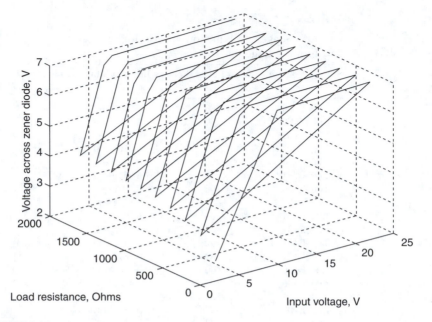

FIGURE 5.15
Output voltage vs. source voltage and load resistance.

The 3-D plot is shown in Figure 5.15.

5.4 Peak Detector

A peak detector is a circuit that can be used to detect the peak value of an input signal. A peak detector can also be used as a demodulator to detect the audio signal in an amplitude-modulated (AM) signal. A simple peak detector circuit is shown in Figure 5.16.

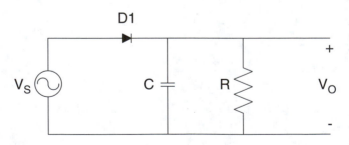

FIGURE 5.16
Peak detector circuit.

The peak detector circuit is simply a half-wave rectifier circuit with a capacitor connected across the load resistor. The operation of the circuit will be described with the assumption that the source voltage is a sinusoidal voltage, $V_m \sin(2\pi f_0 t)$, with amplitude greater than 0.6 V. During the first quarter-cycle, the input voltage will increase and the capacitor will be charged to the input voltage. At time $t = 1/4f_0$, where f_0 is the frequency of the sinusoidal input, the input voltage reaches its maximum value of V_m and the capacitor will be charged to that maximum of V_m.

When time $t > 1/4f_0$, the input starts to decrease, the capacitor will discharge through the resistance R. If we define $t_1 = 1/4f_0$, the time when the capacitor is charged to the maximum value of the input voltage, the discharge of the capacitor is given as

$$v_0(t) = V_m e^{-(t-t_1)/RC} \tag{5.14}$$

The time constant RC should be properly selected to allow the output voltage to approximately represent the peak input signal, within a reasonable error. The following example demonstrates the effects of the time constant.

Example 5.8   Demodulation of AM Wave Using Peak Detection
The peak detector circuit shown in Figure 5.16 is used to detect an amplitude-modulated wave given as

$$v_S(t) = 10\left[1 + 0.5\cos(2\pi f_m t)\right]\cos(2\pi f_c t) \tag{5.15}$$

where
f_c = carrier frequency, in hertz
f_m = modulating frequency, in hertz

If $f_c = 0.2$ MHz, $f_m = 15$KHz, C = 20 nF, RM = 100 Ω, and diode D1 is D1N916, determine the mean square error between the envelope obtained for RL = 1 KΩ and that obtained for the following values of RL: 3, 5, 7, 9, and 11 KΩ.

Solution
The modulated wave can be expressed as

$$v_S(t) = 10\cos(2\pi f_c t) + 5\cos(2\pi f_m t)\cos(2\pi f_c t)$$
$$= 10\cos(2\pi f_c t) + 2.5\cos\left[2\pi(f_c + f_m)t\right] + 2.5\cos\left[2\pi(f_c - f_m)t\right] \tag{5.16}$$

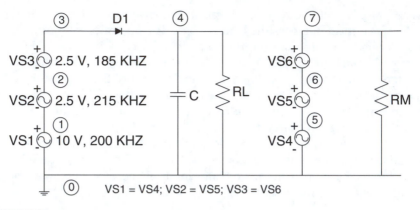

FIGURE 5.17
Demodulator circuit.

The PSPICE program uses sine functions only, so we convert the cosine into sine terms using the trigonometric identity

$$\sin(\omega t + 90°) = \cos(\omega t) \tag{5.17}$$

Equation (5.16) can be rewritten as

$$v_S(t) = 10\sin(2\pi f_c t + 90°) + 2.5\sin\left[2\pi(f_c + f_m)t + 90°\right]$$
$$+ 2.5\sin\left[2\pi(f_c - f_m)t + 90°\right] \tag{5.18}$$

The frequency content of the modulated wave is

f_c = 200 KHz

$f_c + f_m$ = 215 KHz

$f_c - f_m$ = 185 KHz

The demodulator circuit for PSPICE simulation is shown in Figure 5.17.

PSPICE program:

```
AM DEMODULATOR
VS1 1 0 SIN(0 10 200KHZ 0 90)
VS2   2 1   SIN(0 2.5 215KHZ 0 90)
VS3   3 2   SIN(0 2.5 185KHZ 0 90)
D1    3 4   D1N916
```

                                                                    (continued)

```
.MODEL D1N916  D(IS=0.1P RS=8 CJO=1P TT=12N BV=100 IBV=0.1P)
C    4 0 20E-9
RL   4 0 RMOD 1
.MODEL RMOD RES(R=1)
.STEP RES RMOD(R) 1K 11K 2K
VS4 5 0 SIN(0 10 200KHZ 0 90)
VS5 6 5 SIN(0 2.5 215KHZ 0 90)
VS6 7 6 SIN(0 2.5 185KHZ 0 90)
RM   7 0 100
.TRAN 2US 150US
.PRINT TRAN V(4) V(7)
.PROBE V(4) V(7)
.END
```

Figure 5.18 shows the modulated wave and the envelope for an RL of 1 KΩ.

PSPICE partial results for RL = 1 KΩ and 11 KΩ are shown in Tables 5.9 and 5.10, respectively. The complete results for RL = 1, 3, 5, 7, 9, and 11 KΩ can be found in files ex5_8aps.dat, ex5_8bps.dat, ex5_8cps.dat, ex5_8dps.dat, ex5_8eps.dat and ex5_8fps.dat, respectively.

MATLAB is used to determine the mean squared value between the envelope for RL = 1 KΩ and those of the other values of RL.

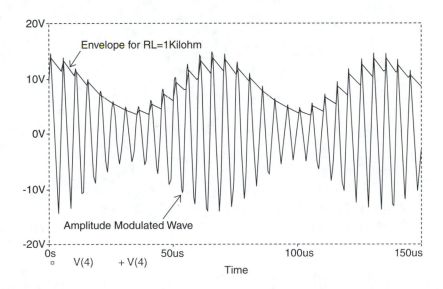

FIGURE 5.18

Amplitude-modulated wave and its envelope for load resistance of 1 KΩ.

TABLE 5.9

Output Voltage for Load
Resistance of 1 KΩ

Time (s)	Output Voltage for RL = 1KΩ (V)
1.000E–05	1.100E+01
2.000E–05	7.645E+00
3.000E–05	4.664E+00
4.000E–05	3.670E+00
5.000E–05	6.168E+00
6.000E–05	9.826E+00
7.000E–05	1.156E+01
8.000E–05	1.001E+01
9.000E–05	6.550E+00
1.000E–04	3.973E+00

TABLE 5.10

Output Voltage for Load
Resistance of 11 KΩ

Time (s)	Output Voltage for RL = 11 KΩ (V)
1.000E–05	1.351E+01
2.000E–05	1.291E+01
3.000E–05	1.234E+01
4.000E–05	1.179E+01
5.000E–05	1.126E+01
6.000E–05	1.143E+01
7.000E–05	1.380E+01
8.000E–05	1.319E+01
9.000E–05	1.260E+01
1.000E–04	1.204E+01

MATLAB script:

```
% Demodulator circuit
%
% Read Data from PSPICE simulations
load 'ex5_8aps.dat' -ascii;
load 'ex5_8bps.dat' -ascii;
load 'ex5_8cps.dat' -ascii;
load 'ex5_8dps.dat' -ascii;
```

(continued)

```
load 'ex5_8eps.dat' -ascii;
load 'ex5_8fps.dat' -ascii;
v1 = ex5_8aps(:,2);
v2 = ex5_8bps(:,2);
v3 = ex5_8cps(:,2);
v4 = ex5_8dps(:,2);
v5 = ex5_8eps(:,2);
v6 = ex5_8fps(:,2);
n = length(v1);   % Number of data points
ms1 = 0;
ms2 = 0;
ms3 = 0;
ms4 = 0;
ms5 = 0;
% Calculate squared error
for i = 1:n
    mse1 = ms1 + (v2(i) - v1(i))^2;
    mse2 = ms2 + (v3(i) - v1(i))^2;
    mse3 = ms3 + (v4(i) - v1(i))^2;
    mse4 = ms4 + (v5(i) - v1(i))^2;
mse5 = ms5 + (v6(i) - v1(i))^2;
end
% Calculate mean squared error
mse(1) = mse1/n;
mse(2) = mse2/n;
mse(3) = mse3/n;
mse(4) = mse4/n;
mse(5) = mse5/n;
r1 = 3e3:2e3:11e3
plot(r1, mse, r1, mse, 'ob')
title('Mean Squared Error as Function of Load
Resistance')
xlabel('Load Resistance, Ohms')
ylabel('Mean Squared Error')
```

The mean squared error is shown in Figure 5.19. It can be seen from the figure that as the load resistance increases, the mean squared error increases.

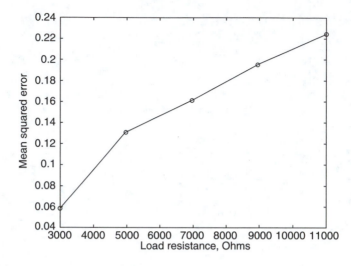

FIGURE 5.19
Mean squared error vs. load resistance.

5.5   Diode Limiters

The general transfer characteristics of a limiter are given by the following expressions:

$$V_0 = kv_{IN} \qquad V_A \le v_{IN} \le V_B$$

$$v_0 = V_H \qquad v_{IN} > V_B \qquad\qquad (5.19)$$

$$V_0 = V_L \qquad v_{IN} < V_A$$

where
   k is a constant
   V_H and V_L are output high and output low voltages, respectively
   v_0 is the output voltage
   v_{IN} is the input voltage

In Equation (5.19), if v_{IN} exceeds the upper threshold V_B, the output is limited to the voltage V_H. In addition, if the input voltage v_{IN} is less than the lower threshold voltage V_A, the output is limited to V_L. Equation (5.19) shows the describing relationships for a double limiter, where the output can be limited to two voltages V_H and V_L. If the output voltage is limited to one voltage, either V_H or V_L then the circuit is a single limiter.

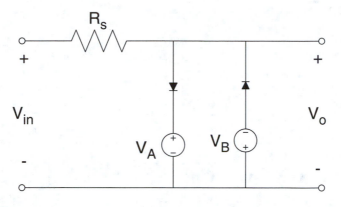

FIGURE 5.20
Double limiter.

In Equation (5.19), V_H and V_L are constant and are independent of the input voltage v_{IN}. Circuits that have such characteristics are termed hard limiters. If V_H and V_L are not constant but are linearly related to the input voltage, then we have soft limiting. A limiter circuit is shown in Figure 5.20.

The transfer characteristics of a double limiter are given as

$$V_0 = V_A + V_D \qquad \text{for} \quad V_0 \geq V_A + V_D$$
$$V_0 = V_{IN} \qquad \text{for} \quad -\left(V_B + V_D\right) \leq V_0 \leq V_A + V_D \qquad (5.20)$$
$$V_0 = -\left(V_B + V_D\right) \quad \text{for} \quad V_0 \leq -\left(V_B + V_D\right)$$

where V_D is the diode drop.

The following example explores a diode limiter circuit.

Example 5.9 Limiter Characteristics

For the precision bipolarity limiter shown in Figure 5.21, VCC = 15 V, VEE = −15 V, R1 = R2 = RA = 10 KΩ. The input signal ranges from −13 V to +13 V and the diodes are D1N916. Determine the minimum value of the output voltage, the maximum value, and the dc transfer characteristic. What is the proportionality constant of the transfer characteristics where there is no limiting?

Solution

PSPICE is used to do the circuit simulation.

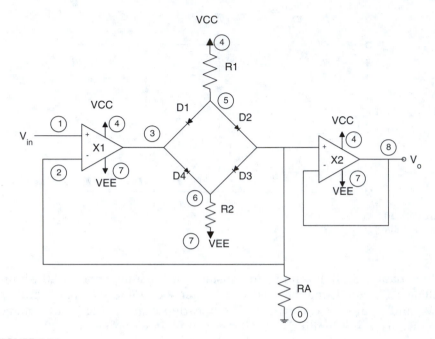

FIGURE 5.21
Precision bipolarity limiter.

PSPICE program:

```
LIMITER CIRCUIT
* CIRCUIT DESCRIPTION
VIN    1    0    DC    0V;
VCC    4    0    DC    15V; 15V POWER SUPPLY
VEE    7    0    DC    -15V; -15V POWER SUPPLY
X1     1    2    4    7    3    UA741;   UA741 OP-AMP
* +INPUT; -INPUT; +VCC; -VEE; OUTPUT; CONNECTIONS FOR OP
AMP UA741
R1     4    5    10K
D1     5    3    D1N916
D2     5    2    D1N916
D3     2    6    D1N916
D4     3    6    D1N916
.MODEL D1N916 D(IS=0.1P RS=8 CJO=1P TT=12N BV=100 IBV=0.1P)
```

(continued)

```
R2      6       7       10K
X2      2       8       4       7   8   UA741;   UA741 OP-AMP
* +INPUT; -INPUT; +VCC; -VEE; OUTPUT; CONNECTIONS FOR OP
AMP UA741
RA      2       0       10K
** ANALYSIS TO BE DONE**
* SWEEP THE INPUT VOLTAGE FROM -12V TO +12 V IN 0.2V
INCREMENTS
.DC   VIN   -13   13      0.5
** OUTPUT REQUESTED
.PRINT DC V(8)
.PROBE V(8)
.LIB NOM.LIB;
* UA741 OP AMP MODEL IN PSPICE LIBRARY FILE NOM.LIB
.END
```

Partial results from the SPICE simulation are shown in Table 5.11. The complete results can be found in file ex5_9ps.dat.

MATLAB is used to perform the data analysis and also to plot the transfer characteristics.

TABLE 5.11

Output vs. Input Voltages of a Limiter

Input Voltage (V)	Output Voltage (V)
−1.300E+01	−7.204E+00
−1.100E+01	−7.204E+00
−9.000E+00	−7.204E+00
−7.000E+00	−6.999E+00
−5.000E+00	−5.000E+00
−3.000E+00	−3.000E+00
−1.000E+00	−9.999E−01
1.000E+00	1.000E+00
3.000E+00	3.000E+00
5.000E+00	5.000E+00
7.000E+00	7.000E+00
9.000E+00	7.203E+00
1.100E+01	7.203E+00
1.300E+01	7.203E+00

MATLAB script:

```
% Limiter Circuit
%
% Read data from file
load 'ex5_9ps.dat' -ascii;
vin = ex5_9ps(:,1);
vout = ex5_9ps(:,2);
% Obtain minimum and maximum value of output
vmin = min(vout); % minimum value of output
vmax = max(vout); % maximum value of output
% Obtain proportionality constant of the non-limiting
region
n = length(vin); % size of data points
% Calculate slopes
for i = 1:n-1
slope(i) = (vout(i+1) - vout(i))/(vin(i+1)-vin(i));
end
kprop = max(slope); % proportionality constant
% Plot the transfer characteristics
plot(vin, vout)
title('Transfer Characteristics of a Limiter')
xlabel('Input Voltage, V')
ylabel('Output Voltage, V')
% Print the results
fprintf('Maximum Output Voltage is %10.4e V\n', vmax)
fprintf('Minimum Output Voltage is %10.4e V\n', vmin)
fprintf('Proportionality Constant is %10.5e V\n', kprop)
```

The transfer characteristics are shown in Figure 5.22.
The following results were obtained from the MATLAB program:

Maximum Output Voltage is 7.2030E+000 V
Minimum Output Voltage is –7.2040E+000 V
Proportionality Constant is 1.00020E+000 V

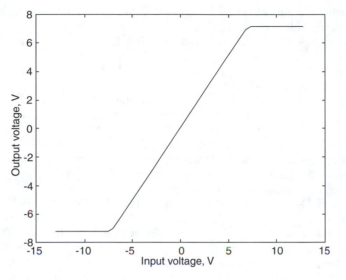

FIGURE 5.22
Transfer characteristics of a limiter.

Bibliography

1. Al-Hashimi, Bashir, *The Art of Simulation Using PSPICE, Analog, and Digital,* CRC Press, Boca Raton, FL, 1994.
2. Attia, J.O., *Electronics and Circuit Analysis Using MATLAB,* CRC Press, Boca Raton, FL, 1999.
3. Biran, A. and Breiner, M., *MATLAB for Engineers,* Addison-Wesley, Reading, MA, 1995.
4. Chapman, S. J., *MATLAB Programming for Engineers,* Brook, Cole Thompson Learning, Pacific Grove, CA, 2000.
5. Connelly, J. Alvin and Choi, Pyung, *Macromodeling with SPICE,* Prentice-Hall, Englewood Cliffs, NJ, 1992.
6. Derenzo, S.E., Interfacing: *A Laboratory Approach Using the Micrcomputer for Instrumentation, Data Analysis and Control,* Prentice-Hall, Englewood Cliffs, NJ, 1990.
7. Etter, D.M., *Engineering Problem Solving with MATLAB,* 2nd edition, Prentice-Hall, Upper Saddle River, NJ, 1997.
8. Etter, D.M., Kuncicky, D.C., and Hull, D., *Introduction to MATLAB 6,* Prentice-Hall, Upper Saddle River, NJ, 2002.
9. Fenical, L. H., *PSPICE: A Tutorial,* Prentice-Hall, Englewood Cliffs, NJ, 1992.
10. Hart, Daniel W., Introducing Fourier Series Using PSPICE Computer Simulation, *Computers in Education, Division of ASEE,* Vol. III, No. 2, pp. 46–51, April–June 1993.
11. Keown, John, *PSPICE and Circuit Analysis,* Maxwell MacMillan International Publishing Group, New York, 1991.

12. Kielkowski, Ron M., *Inside SPICE, Overcoming the Obstacles of Circuit Simulation*, McGraw-Hill, New York, 1994.

13. Lamey, Robert, *The Illustrated Guide to PSPICE*, Delmar Publishers, Albany, NY, 1995.

14. Nilsson, James W. and Riedel, Susan A., *Introduction to PSPICE*, Addison-Wesley, Reading, MA, 1993.

15. OrCAD PSPICE A/D, Users' Guide, November 1998.

16. Prigozy, Stephen, Novel Applications of PSPICE in Engineering, *IEEE Trans. Education*, 32(1), 35–38, Feb. 1989.

17. Rashid, Mohammad H., *Microelectronic Circuits, Analysis and Design*, PWS Publishing Company, Boston, MA, 1999.

18. Rashid, Mohammad H., *SPICE for Circuits and Electronics Using PSPICE*, Prentice-Hall, Englewood Cliffs, NJ, 1990.

19. Roberts, Gordon W. and Sedra, Adel S., *SPICE for Microelectronic Circuits*, Saunders College Publishing, Fort Worth, TX, 1992.

20. Sedra, A.S. and Smith, K.C., Microelectronic Circuits, 4th edition, Oxford University Press, New York, 1998.

21. Sigmor, K., *MATLAB Primer*, 4th edition, CRC Press, Boca Raton, FL, 1998.

22. Tuinenga, Paul W., *SPICE, A Guide to Circuit Simulations and Analysis Using PSPICE*, Prentice-Hall, Englewood Cliffs, NJ, 1988.

23. Using MATLAB, The Language of Technical Computing, Computation, Visualization, and Programming, Version 6, MathWorks, Inc., 2000.

24. Vladimirescu, Andrei, *The SPICE Book*, John Wiley & Sons, New York, 1994.

Problems

5.1 A forward-biased diode has the corresponding voltage and current shown in Table P5.1. (a) Determine the equation of best fit. (b) For the voltage of 0.64 V, what is the diode current?

TABLE P5.1

Voltage vs. Current of a Diode

Forward–Biased Voltage (V)	Forward Current (A)
0.1	1.33E–13
0.2	1.79E–12
0.3	24.02E–12
0.4	0.321E–9
0.5	4.31E–9
0.6	57.69E–9
0.7	7.72E–7

5.2 For Example 5.2, plot the diode current as a function of temperature. Determine the equation of best fit between the diode current and temperature.

5.3 For the diode circuit shown in Figure P5.3, R = 10 KΩ and D1 is D1N916. If the voltage VDC increases from 0.3 V to 2.1 V in increments of 0.2 V, determine the diode dynamic resistance and the diode voltage for the various values of the input voltage. Plot the dynamic resistance vs. diode voltage.

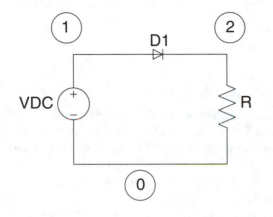

FIGURE P5.3
Diode circuit.

5.4 For the battery charging circuit shown in Figure 5.5, $v_s(t) = 18\sin(120\pi t)$, R = 100 Ω, and VB = 12 V. Determine the conduction angle. Assume that the diode is D1N4448.

5.5 For the battery charging circuit shown in Figure 5.5, $v_s(t) = 18\sin(120\pi t)$, R = 100 Ω, and VB increases from 10.5 to 12.0 V. Plot the average current flowing through the diode as a function of the battery voltage. Assume that the diode is D1N4448.

5.6 For the half-wave rectifier with smoothing circuit shown in Figure P5.6, diode D1 is D1N4009, LP = 1 H, LS = 100 mH, RS = 5 Ω, and RL = 10 KΩ. (a) If C = 10 μF, plot the diode current with respect to time. (b) If capacitor C takes on the values of 1 μF 10 μF, 100 μF, and 500μF, (i) plot the rms value of the output voltage with respect to the capacitance values, and (ii) plot the average diode current with respect to the capacitance values. Assume that the coefficient of coupling of the transformer is 0.999.

5.7 For the full-wave rectifier with the smoothing circuit shown in Figure 5.9, diodes D1, D2, D3, and D4 are D1N914 and C = 10 μF. Determine the peak diode current as RL takes on the following values: 10 KΩ, 30 KΩ, 50 KΩ, 70 KΩ, 90 KΩ, and 110 KΩ. Plot the peak diode current as a function of RL.

5.8 For Figure 5.12, assume that VS = 18 V, D1 is D1N4742, RL = 100 Ω, and RS = 2 Ω. Find the voltage across the diode as the temperature various from 0°C to 100°C. Plot the diode voltage as a function of temperature.

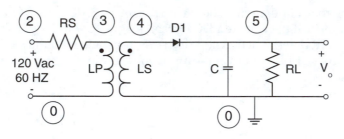

FIGURE P5.6
Rectifier circuit.

5.9 In the Zener diode shunt voltage regulator shown in Figure 5.12, the source voltage is 20 V and the Zener diode is D1N4742. The load resistance varies from 1 KΩ to 20 KΩ and the source resistance RS is from 20 Ω to 100 Ω. Plot the output voltage as a function of both the load resistance and the source resistance.

5.10 In Figure 5.12, VS varies from 8 V to 20 V and RL = 2 KΩ. The resistance RS varies from 20 Ω to 120 Ω . The Zener diode is D1N4742. Plot the output voltage as a function of both the source voltage VS and source resistance RS.

5.11 In the peak detector circuit shown in Figure 5.16, $v_s(t) = 15\sin(360\pi t)$ V, C = 0.01 μF, and the diode is D1N4009. As the load resistance takes on the values of 1, 3, 5, 7, and 9 KΩ. (a) determine the peak diode currents as a function of load resistance, and (b) obtain the mean value of the diode current for one complete period of the input signal.

5.12 In Figure 5.21, R1 = R2 = RA = 10 KΩ and $V_{in}(t) = 10 \sin(240\pi t)$ V. X1 and X2 are 741 op amps. If VCC = |VEE| = V_K and V_K is changed from 11 V to 16 V, what will be the minimum and maximum values of the output voltage? Assume that the diodes are D1N916.

5.13 For the limiter circuit shown in Figure P5.13, VCC = 15 V, VEE = –15 V, R1 = 2 KΩ, R2 = 4 KΩ, and the diodes are D1N754. Determine the peak and minimum values of the output voltage if the sinusoidal voltage of 20 sin(960πt) is applied at the input. Plot the output voltage. Assume a UA741 op amp.

5.14 In the back-to-back diode limiter circuit shown in Figure P5.14, R1 = 1 KΩ and the diodes D1 and D2 are D1N914. Determine the output voltage if the input signal is sinusoidal wave $V_{in}(t) = 2 \sin(960\pi t)$ V. Calculate the percent distortion for the 2nd, 3rd, and 4th harmonics.

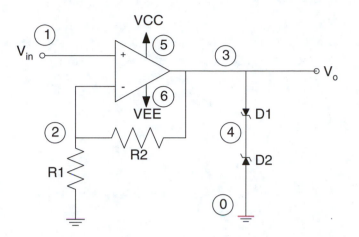

FIGURE P5.13
Limiter circuit.

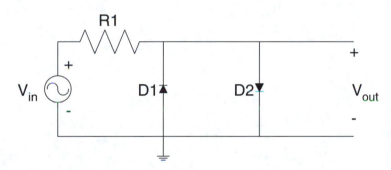

FIGURE P5.14
Back-to-back diode limiter.

6

Operational Amplifier

Operational amplifiers (op amps) are highly versatile. They can be used to perform mathematical operations such as addition, subtraction, multiplication, integration, and differentiation. Several electronic circuits, such as amplifiers, filters, oscillators, and flip-flops use operational amplifiers as an integral element. This chapter discusses the properties of op amps. The non-ideal characteristics will be explored. The analysis of op amp filters and comparator circuits will be done using both PSPICE and MATLAB programs.

6.1   Inverting and Non-inverting Configurations

The op amp, from the signal point of view, is a three-terminal device. The ideal op amp has infinite resistance, zero output resistance, zero offset voltage, infinite frequency response, infinite common-mode rejection ratio, and infinite open-loop gain.

A practical op amp will have large but finite open-loop gain in the range from 10^5 to $10^9\,\Omega$. It also has a very large input resistance, from 10^6 to $10^{10}\,\Omega$. The output resistance might be in the range of 50 to 125 Ω. The offset voltage is small but finite, and the frequency response will deviate considerably from the infinite frequency response of the ideal op amp.

6.1.1   Inverting Configuration

One basic configuration of the op amp is the inverted closed-loop configuration shown in Figure 6.1.

It can be shown that

$$\frac{V_O}{V_{IN}} = -\frac{Z_2}{Z_1}$$

(6.1)

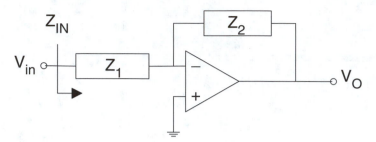

FIGURE 6.1
Inverting configurations.

and the input impedance is

$$Z_{IN} = Z_1 \tag{6.2}$$

Case 1: **Inverting amplifier**
If $Z_1 = R_1$ and $Z_2 = R_2$, we have an inverting amplifier and the closed-loop gain is

$$\frac{V_0}{V_{IN}} = -\frac{R_2}{R_1} \tag{6.3}$$

Case 2: **Miller integrator**
If $Z_1 = R_1$ and $Z_2 = 1/j\omega C$, we obtain an integrator termed the Miller integrator. The closed-loop gain of the integrator is

$$\frac{V_0}{V_{IN}} = \frac{-1}{j\omega CR_1} \tag{6.4}$$

In the time domain, the above expression becomes

$$V_0(t) = -\frac{1}{CR_1}\int_0^t V_{IN}(t)dt + V_0(0) \tag{6.5}$$

Case 3: **Differentiator circuit**
If $Z_1 = 1/j\omega C$ and $Z_2 = R$, we obtain a differentiator circuit. The closed-loop gain of a differentiator is

$$\frac{V_0}{V_{IN}} = -j\omega CR \tag{6.6}$$

In the time domain, the above expression becomes

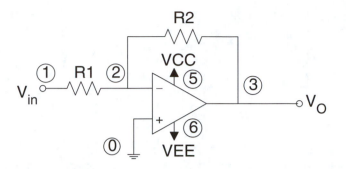

FIGURE 6.2
Inverting amplifier.

$$V_o(t) = -CR\frac{dV_{IN}(t)}{dt} \qquad (6.7)$$

The following example explores the behavior of an inverting amplifier.

Example 6.1 DC Transfer Characteristics of an Inverting Amplifier

For the inverting amplifier shown in Figure 6.2, VCC = 15 V, VEE = −15 V, R1 = 2 KΩ, and R2 = 5 KΩ.

1. Determine the maximum and minimum output voltages.
2. What is the gain in the region that provides linear amplification?
3. Determine the range of input voltage that provides linear amplification.

Solution

PSPICE is used to obtain the transfer characteristics. MATLAB is then employed for data processing.

PSPICE program:

```
DC TRANSFER CHARACTERISTICS
VIN 1   0   DC 0.5V
R1   1   2   1E3
R2   2   3   5E3
VCC 5   0   DC 15V; POWER SUPPLY
VEE 6   0   DC -15V; POWER SUPPLY
X1   0   2   5   6   3 UA741; UA741 OP AMP
*  +INPUT; -INPUT; +VCC; -VEE; OUTPUT; CONNECTIONS FOR UA741
```

(continued)

```
** ANALYSIS TO BE DONE
.DC VIN -14 +14 0.5V
.LIB NOM.LIB
* UA741 OP AMP MODEL IN PSPICE LIBRARY FILE NOM.LIB
** OUTPUT
.PRINT DC V(3)
.END
```

Partial results from the PSPICE simulation are shown in Table 6.1. The complete results are in file ex6_1ps.dat.

TABLE 6.1

Input and Output Voltages of
Inverting Amplifier of Figure 6.2

Input Voltage (V)	Output Voltage (V)
−1.400E+01	1.461E+01
−5.000E+00	1.461E+01
−3.000E+00	1.460E+01
−1.000E+00	5.000E+00
0.000E+00	5.143E−04
1.000E+00	−4.999E+00
3.000E+00	−1.460E+01
5.000E+00	−1.461E+01
1.400E+01	−1.461E+01

MATLAB is used for processing the input/output data of the amplifier.

MATLAB script:

```
% Analysis of input/output data using MATLAB
% Read data using load command
load 'ex6_1ps.dat'
vin = ex6_1ps(:,1);
vo = ex6_1ps(:,2);
% Plot transfer characteristics
plot(vin, vo)
xlabel('Input Voltage, V')
ylabel('Output Voltage, V')
```

(continued)

```
title('Transfer Characteristics')
vo_max = max(vo);   % maximum value of output
vo_min = min(vo);   % minimum value of output
% calculation of gain
m = length(vin);
m2 = fix(m/2);
gain = (vo(m2 + 1) - vo(m2 - 1))/(vin(m2 + 1) - vin(m2 - 1));
% range of input voltage with linear amp
vin_min = vo_min/gain; % maximum input voltage
vin_max = vo_max/gain;  % minimum input voltage
% print out
fprintf ('Maximum Output Voltage is %10.4eV\n', vo_max)
fprintf('Minimum Output Voltage is %10.4eV\n', vo_min)
fprintf('Gain is %10.5e\n', gain)
fprintf('Minimum input voltage for Linear Amplification
is %10.5e\n,', vin_max)
fprintf('Maximum input voltage for Linear Amplification
is %10.5e\n,', vin_min)
```

The MATLAB results are

Maximum Output Voltage is 1.4610E+001V
Minimum Output Voltage is –1.4610E+001V
Gain is –4.99949E+000
Minimum input voltage for Linear Amplification is –2.92230E+000
Maximum input voltage for Linear Amplification is 2.92230E+000

The transfer characteristics are shown in Figure 6.3.

6.1.2   Non-inverting Configuration

Another basic configuration of the op amp is the non-inverting configuration. It is shown in Figure 6.4.

It can easily be shown that the input and output voltages are related by

$$\frac{V_O}{V_{IN}} = 1 + \frac{Z_2}{Z_1} \qquad (6.8)$$

If $Z_1 = R_1$ and $Z_2 = R_2$, Figure 6.4 becomes a voltage follower with gain. The gain is

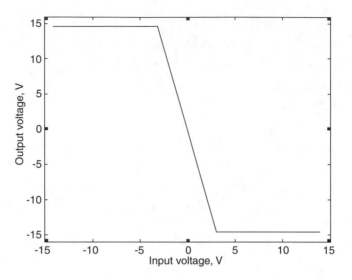

FIGURE 6.3
DC transfer characteristics of an inverting amplifier.

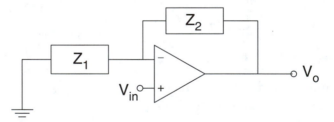

FIGURE 6.4
Non-inverting configuration.

$$\frac{V_O}{V_{IN}} = 1 + \frac{R_2}{R_1} \tag{6.9}$$

The non-inverting configuration has very high input resistance and the output voltage is in phase with the input voltage. The following example explores the frequency response of a non-inverting amplifier.

Example 6.2 Unity Gain Bandwidth of an Op Amp

For the non-inverting amplifier shown in Figure 6.5, VCC = 15 V, VEE = −15 V, R1 = 1 KΩ, and R2 = 9 KΩ. Determine the frequency response of the amplifier. Calculate the (a) 3-dB frequency, (b) unity-gain bandwidth, and (c) gain at midband. Assume that X1 is a UA741 op amp.

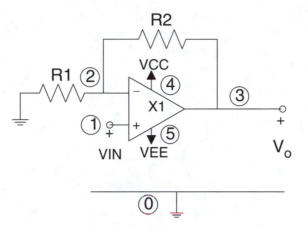

FIGURE 6.5
Non-inverting op amp.

Solution

PSPICE is used to obtain the frequency response. MATLAB is then employed to calculate the 3-dB frequency, unity-gain bandwidth, and the gain at midband.

PSPICE program:

```
FREQUENCY RESPONSE OF NON-INVERTING AMP
VIN 1  0  AC 1V 0
R1  2  0  1000
R2  2  3  9000
X1  1  2  4  5  3 UA741; UA741 OP-AMP
* +INPUT; -INPUT; +VCC; -VEE; OUTPUT; CONNECTIONS FOR UA741
VCC 4  0  DC 15V;   15 V POWER SUPPLY
VEE 5  0  DC -15V;  -15 V POWER SUPPLY
* ANALYSIS TO BE DONE
.AC DEC 5 0.1HZ 100MEGHZ
.LIB NOM.LIB;
* UA741 OP AMP MODEL IN PSPICE LIBRARY FILE NOM.LIB
* OUTPUT
.PRINT AC VDB(3)
.END
```

Partial results from the PSPICE simulation are shown in Table 6.2. The complete results can be found in file ex6_2ps.dat.

TABLE 6.2

Frequency Response of a
Non-inverting Amplifier

Frequency (Hz)	Gain (dB)
1.000E–01	2.000E+01
3.981E–01	2.000E+01
1.000E+00	2.000E+01
3.981E+00	2.000E+01
1.000E+01	2.000E+01
3.981E+01	2.000E+01
1.000E+02	2.000E+01
3.981E+02	2.000E+01
1.000E+03	2.000E+01
1.000E+04	1.996E+01
3.981E+04	1.942E+01
1.000E+05	1.721E+01
3.981E+05	7.931E+00
1.000E+06	–9.620E–01
3.981E+06	–1.997E+01
1.000E+07	–3.541E+01
3.981E+07	–5.934E+01
1.000E+08	–7.555E+01

The MATLAB program for analyzing the PSPICE results follows.

MATLAB script:

```
% Frequency Response of a Non-inverting Amplifier
load 'ex6_2ps.dat' -ascii;
freq = ex6_2ps(:,1);
gain = ex6_2ps(:,2);
% Plot the frequency response
plot(freq, gain)
xlabel('Frequency, Hz')
ylabel('Gain, dB')
title('Frequency Response of a Non-inverting Amplifier')
% calculations
g_mb = gain(1); % midband gain
% Unity gain bandwidth in frequency at which gain is unity
% Cut-off frequency is frequency at which gain is
approximately
% 3dB less than the midband gain
```

(continued)

```
m = length(freq);  % number of data points
for i = 1:m
    g1(i) = abs(gain(i) - g_mb +3);
    g2(i) = abs(gain(i));
end
% cut-off frequency
[f6, n3dB] = min(g1);
freq_3dB = freq(n3dB);   % 3dB frequency
% Unity Gain Bandwidth
[f7, n0dB] = min(g2);
freq_0dB = freq(n0dB);   % unity gain
% print results
fprintf('Midband Gain is %10.4e dB\n', g_mb)
fprintf('3dB frequency is %10.4e Hz\n', freq_3dB)
fprintf('Unity Gain Bandwidth is %10.4e Hz\n', freq_0dB)
```

The MATLAB results are

Midband Gain is 2.0000E+001 dB
3dB frequency is 1.0000E+005 Hz
Unity Gain Bandwidth is 1.0000E+006 Hz

The frequency response is shown in Figure 6.6.

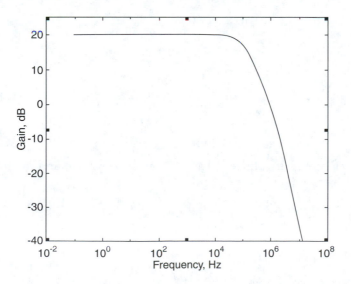

FIGURE 6.6
Frequency response of a non-inverting amplifier.

6.2   Slew Rate and Full-Power Bandwidth

Slew rate (SR) is a measure of the maximum possible rate of change of the output voltage of an op amp. Mathematically, it is defined as

$$SR = \frac{dV_O}{dt}\bigg|_{max} \tag{6.10}$$

Slew rate becomes important when an output signal of an op amp must follow a signal at the input of the op amp that is large in amplitude and rapidly changing with time. If the slew rate is lower than the rate of change of the input signal, then the output will be distorted. However, if the slew rate is higher than the rate of change of the input signal, then no distortion occurs and the input and output of the op amp circuit will have similar wave shapes.

The full power bandwidth f_m is the frequency at which a sinusoidal rated output signal begins to show distortion due to slew rate limiting. A sinusoidal input voltage given by

$$V_i(t) = V_m \sin(2\pi f_m t) \tag{6.11}$$

If the above signal is applied to a voltage follower with unity gain, then the output rate of change is given as

$$\frac{dV_O(t)}{dt} = \frac{dV_i(t)}{dt} = 2\pi f_m \cos(2\pi f_m t) \tag{6.12}$$

From Equations (6.10) and (6.12),

$$SR = \frac{dV_O}{dt}\bigg|_{max} = 2\pi f_m V_m \tag{6.13}$$

If the rated output voltage is $V_{0,rated}$, then the slew rate and the full power bandwidth are related by

$$2\pi f_m V_{0,rated} = SR \tag{6.14}$$

Thus,

$$f_m = \frac{SR}{2\pi V_{0,rated}} \tag{6.15}$$

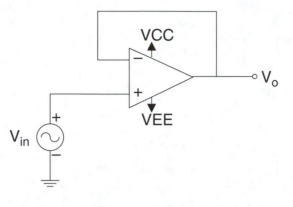

FIGURE 6.7
Voltage follower.

From Equation (6.14), the full-power bandwidth can be traded for output rated voltage. Thus, if the output rated voltage is reduced, the full-power bandwidth increases. The following example explores slew rate and signal distortion.

Example 6.3   Slew Rate and Full-Power Bandwidth

For the voltage follower circuit shown in Figure 6.7, VCC = 15 V and VEE = −15 V. Determine the slew rate if the op amp is a UA741. If the rated output voltage varies from 8 V to 14.5 V, determine the full-power bandwidth. Plot the full-power bandwidth as a function of the rated output voltage.

Solution

For the slew rate calculation, a pulse waveform is applied at the input and the slope of the rising edge of the output waveform is calculated. Using Figure 6.8, the PSPICE program is used to obtain the output waveform.

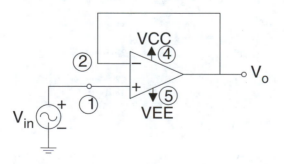

FIGURE 6.8
Slew rate calculation.

PSPICE program:

```
SLEW RATE
* SLEW RATE CALCULATION
VIN1  1 0 PULSE (0 10 0 10NS 10NS 10U 20U)
VCC 4 0 DC 15V
VEE 5 0 DC -15V
X1  1 2 4 5 2 UA741; UA741 OP AMP
* +INPUT; -INPUT; +VCC; -VEE; OUTPUT; CONNECTIONS FOR UA741
.LIB NOM.LIB;
* UA741 OP AMP MODEL IN PSPICE LIBRARY FILE NOM.LIB
* ANALYSIS TO BE DONE
.TRAN 0.5U 40U
.PRINT TRAN V(1) V(2)
.PROBE V(1) V(2)
.END
```

The partial results of the output voltage obtained from PSPICE are shown in Table 6.3. The results can be found in file ex6_3ps.dat.

TABLE 6.3

Output Voltage vs. Time

Time (s)	Output Voltage (V)
0.000E+00	1.925E–05
2.000E–06	9.948E–01
4.000E–06	2.038E+00
6.000E–06	3.080E+00
8.000E–06	4.123E+00
1.000E–05	5.166E+00
1.200E–05	4.254E+00
1.400E–05	3.237E+00
1.600E–05	2.226E+00
1.800E–05	1.220E+00
2.000E–05	2.189E–01

MATLAB is used to calculate the slew rate and also to plot the full-power bandwidth as a function of input voltage.

MATLAB script:

```
% Slew rate and full-power bandwidth
% read data
load 'ex6_3ps.dat' -ascii;
t = ex6_3ps(:,1);
vo = ex6_3ps(:,3);
% slew rate calculation
nt = length(t); % data points
% calculate derivative with MATLAB function diff
dvo = diff(vo)./diff(t);  % derivative of output with
respect to % time
% find max of the derivative
sr = max(dvo);
vo_rated = 8.0:0.5:14.5;
ko = length(vo_rated);
for i = 1:ko
    fm(i) = sr/(2*pi*vo_rated(i));
end
% plot
subplot(211), plot(t,vo)
xlabel('Time,s')
ylabel('Output Voltage')
title('Output Voltage and Full-Power Bandwidth')
subplot(212),plot(vo_rated,fm)
xlabel('Rated Output Voltage, V')
ylabel('Full-power Bandwidth')
fprintf('Slew Rate is %10.5e V/s\n',sr)
```

The output voltage as a function of time was used to calculate the slew rate. Equation (6.15) was used to calculate the full-power bandwidth. The output voltage and the full-power bandwidth plots are shown in Figure 6.9. Slew Rate is 5.22200E+005 V/s.

The following example explores the effects of input voltage amplitude and frequency on the output of an operational amplifier.

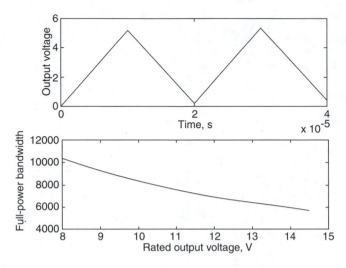

FIGURE 6.9
Ouput voltage (upper) and full-power bandwidth (lower).

Example 6.4 3-D Plot of Output Voltage with Respect to Input Voltage and Frequency

For the voltage follower circuit shown in Figure 6.7, the input signal is a sinusoidal signal given as

$$V_{in}(t) = V_m \sin(2\pi ft) \tag{6.16}$$

If peak voltage V_m varies from 1 V to 11 V and the frequency f varies from 20 KHz to 200 KHz, determine the peak output voltage. Plot the peak output voltage as a function of peak input voltage and frequency.

Solution

PSPICE is used to obtain the output voltage as a function of the changes of the amplitude and frequency of the input sinusoid.

PSPICE program:

```
OUTPUT VOLTAGE AS A FUNCTION OF INPUT VOLTAGE AND FREQUENCY
.PARAM PEAK = 1.0V
.PARAM FREQ = 20KHz
VIN 1 0 SIN(0 {PEAK} {FREQ})
X2  1 2 4 5 2 UA741: U741 Op Amp
*  +INPUT; -INPUT; +VCC; -VEE; OUTPUT; CONNECTIONS FOR UA741
```

(continued)

```
.LIB NOM.LIB;
* UA741 OP AMP MODEL IN PSPICE LIBRARY FILE NOM.LIB
VCC 4 0 DC 15V
VEE 5 0 DC -15V
.STEP PARAM FREQ 20KHz 200KHz 20KHz
.TRAN 0.01US 40US
.PRINT TRAN V(2)
.PROBE V(2)
.END
```

The peak value of the input voltage is changed from 1 V to 11 V in increments of 2 V; the peak output voltage is then obtained for each corresponding value of input voltage and frequency. Partial results from PSPICE simulation are shown in Table 6.4. The complete data can be found in file ex6_4ps.dat.

TABLE 6.4

Output Voltage as a Function of Load
Resistance and Input Voltage

Amplitude of Input Voltage (V)	Frequency of Input Voltage (KHz)	Peak Output Voltage (V)
1	20	1.0
1	40	1.0
1	60	1.0
3	20	3.0
3	40	3.0
3	60	2.57
5	20	5.0
5	40	4.235
5	60	3.205
7	20	6.856
7	40	4.235
7	60	3.205

MATLAB script:

```
% 3D Plot of Output Voltage as a function of input
voltage amplitude
% and frequency.
% Read data
[vin_amp, vin_freq, vout] = textread('ex6_4ps.dat', '%d
%d %f');
```

(continued)

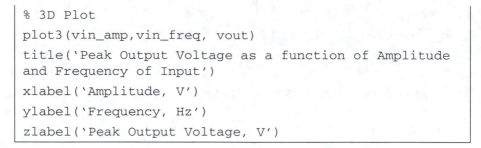

```
% 3D Plot
plot3(vin_amp,vin_freq, vout)
title('Peak Output Voltage as a function of Amplitude
and Frequency of Input')
xlabel('Amplitude, V')
ylabel('Frequency, Hz')
zlabel('Peak Output Voltage, V')
```

The 3-D plot is shown in Figure 6.10.

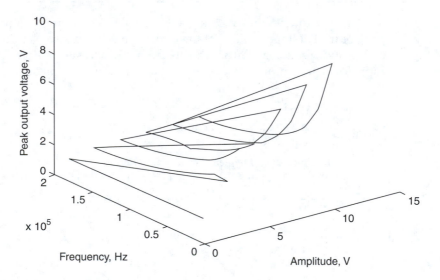

FIGURE 6.10
3-D plot of output voltage.

From Figure 6.10 and Table 6.4, it appears that for small input voltage and frequency, the peak output voltage is almost equal to the amplitude of the input signal. However, with high input voltage and frequency, the peak output voltage is considerably lower than the input amplitude. This is due to slew rate limiting.

6.3 Active Filter Circuits

Electronic filter circuits are circuits that can be used to attenuate particular band(s) of frequency and also pass other band(s) of frequency. The following types of filters are discussed in this section: lowpass, bandpass, highpass, and band-reject. The filters have passband, stopband, and transition band.

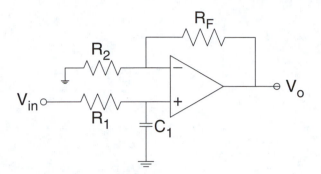

FIGURE 6.11
First-order lowpass filter.

The order of the filter will determine the transition from the passband to the stopband.

6.3.1 Lowpass Filters

A lowpass filter passes low frequencies and attenuates high frequencies. The transfer function of a first-order lowpass filter has the general form

$$H(s) = \frac{k}{s + \omega_0} \tag{6.17}$$

A circuit that can be used to implement a first-order lowpass filter is shown in Figure 6.11.

The voltage transfer function for Figure 6.11 is

$$H(s) = \frac{V_0(s)}{V_{IN}(s)} = \frac{k}{1 + sR_1C_1} \tag{6.18}$$

with dc gain k given as

$$k = 1 + \frac{R_F}{R_2} \tag{6.19}$$

and the cut-off frequency f_0 given as

$$f_0 = \frac{1}{2\pi R_1 C_1} \tag{6.20}$$

The first-order filter exhibits –20 dB/decade roll-off in the stopband. We next consider second-order lowpass filter that has –40 dB/decade roll-off.

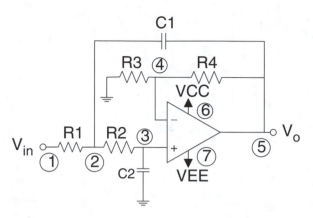

FIGURE 6.12
Sallen-Key lowpass filter.

The second-order lowpass filter can be used as a building block for higher-order filters. The general transfer function for a second-order lowpass filter is

$$H(s) = \frac{k\omega_0^2}{s^2 + \left(\frac{\omega_0}{Q}\right)s + \omega_0^2} \qquad (6.21)$$

where
    ω_0 is the resonant frequency
    Q is the quality factor or figure of merit
    k is the dc gain

The quality factor Q is related to the bandwidth BW and ω_0 by the expression

$$Q = \frac{\omega_0}{BW} = \frac{\omega_0}{\omega_H - \omega_L} \qquad (6.22)$$

where
    ω_H = high cut-off frequency, in rad/s
    ω_L = low cut-off frequency, in rad/s

A circuit that can be used to realize a second-order lowpass filter is the Sallen-Key filter shown in Figure 6.12.

The following example explores the characteristics of the Sallen-Key lowpass filter circuit.

Example 6.5   Sallen-Key Lowpass Filter

The Sallen-Key lowpass filter shown in Figure 6.12 has the following element values: R1 = R2 = 30 KΩ, R3 = 10 KΩ, R4 = 40 KΩ, and C1 = C2 = 0.005 μF.

(a) Determine low frequency gain, (b) find the cut-off frequency, and (c) plot the frequency response.

Solution

PSPICE is used for circuit simulation.

PSPICE program:

```
*SALLEN-KEY LOWPASS FILTER
VIN   1    0    AC    1V
R1    1    2    30K
R2    2    3    30K
R3    4    0    10K
R4    5    4    40K
C1    2    5    0.005UF
C2    3    0    0.005UF
X1    3    4    6    7    5 UA741: UA741 OP AMP
* +INPUT; -INPUT; +VCC; -VEE; OUTPUT; CONNECTIONS FOR UA741
.LIB NOM.LIB;
* UA741 OP AMP MODEL IN PSPICE LIBRARY FILE NOM.LIB
.AC DEC 10 1HZ 100KHZ
.PRINT AC VM(5)
.PROBE V(5)
.END
```

Partial results obtained from the PSPICE simulation are shown in Table 6.5. The complete output data can be found in file ex6_5ps.dat.

TABLE 6.5

Gain vs. Frequency
of a Lowpass Filter

Frequency (Hz)	Gain
1.000E+00	5.009E+00
5.012E+00	5.009E+00
1.000E+01	5.009E+00
5.012E+01	4.998E+00
1.000E+02	4.966E+00
5.012E+02	4.098E+00
1.000E+03	2.644E+00
5.012E+03	2.097E-01
1.000E+04	5.428E-02
5.012E+04	2.176E-03
1.000E+05	9.273E-04

MATLAB is used for the data analysis.

MATLAB script:

```
% Lowpass gain and cut-off frequency
% Read data
load 'ex6_5ps.dat' -ascii;
freq = ex6_5ps(:,1);
vout = ex6_5ps(:,2);
gain_lf = vout(2): % Low frequency gain
gain_cf = 0.707 * gain_lf: % gain at cut-off frequency
tol = 1.0e-3: % tolerance for obtaining cut-off frequency
i = 2: % Initialize the counter
% Use while loop to obtain the cut-off frequency
while (vout(i) - gain_cf) > tol
    i = i +1;
end
m = i;
freq_cf = freq(m): % cut-off frequency
% Print out the results
plot(freq,vout)
xlabel('Frequency, Hz')
ylabel('Gain')
title('Frequency Response of a Lowpass Filter')
fprintf('Low frequency gain is %10.5e\n', gain_lf)
fprintf('Cut-off frequency is %10.5e\n', freq_cf)
```

The results are

Low frequency gain is 5.00900E+000
Cut-off frequency is 7.94300E+002

The frequency response of the lowpass is shown in Figure 6.13.

6.3.2 Highpass Filters

A highpass filter passes high frequencies and attenuates low frequencies. The transfer function of the first-order highpass filter has the general form:

$$H(s) = \frac{ks}{s + \omega_0} \qquad (6.23)$$

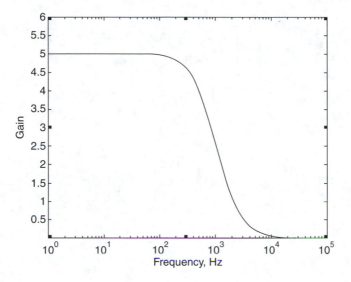

FIGURE 6.13
Frequency response of a lowpass filter.

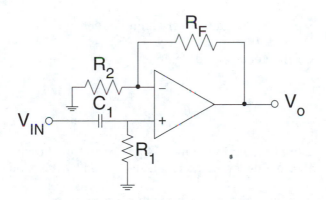

FIGURE 6.14
First-order highpass filter.

The circuit shown in Figure 6.14 can be used to implement the first-order highpass filter. It is basically the same as Figure 6.11, except that the positions of R_1 and C_1 in Figure 6.11 have been interchanged.

For Figure 6.14, the voltage transfer function is

$$H(s) = \frac{V_O}{V_{IN}}(s) = \frac{s}{1 + \frac{1}{R_1 C_1}}\left(1 + \frac{R_F}{R_2}\right) \tag{6.24}$$

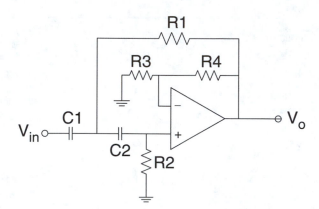

FIGURE 6.15
Sallen-Key highpass filter.

where

$$k = \left(1 + \frac{R_F}{R_2}\right)$$

(6.25)

= gain at very high frequency

and the cut-off frequency at 3 dB gain is

$$f_0 = \frac{1}{2\pi R_1 C_1}$$

(6.26)

Although the filter shown in Figure 6.14 passes all signal frequencies higher than f_0, the high-frequency characteristic is limited by the bandwidth of the op amp.

The second-order highpass filter has the general form

$$H(s) = \frac{ks^2}{s^2 + \left(\frac{\omega}{Q}\right)s + \omega_0^2}$$

(6.27)

where k is the high-frequency gain.

The second-order highpass filter can be obtained from the second-order lowpass filter shown in Figure 6.15. This figure is similar to Figure 6.12 except that the frequency-dominant resistors and capacitors have been interchanged.

The following example will explore the highpass filter characteristics.

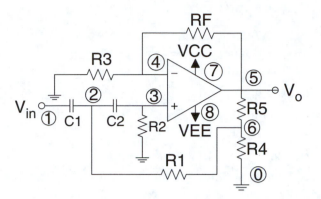

FIGURE 6.16
Modified Sallen-Key highpass filter.

Example 6.6   Highpass Filter

The Sallen-Key highpass filter shown in Figure 6.15 can be modified to obtain various quality factors by including two resistors R4 and R5 at the output. The modified Sallen-Key highpass filter is shown in Figure 6.16. In the latter figure, assume that VCC = 15 V, VEE = –15 V, C1 = C2 = 0.05 µF, R1 = R2 = R3 = R4 = 600 Ω, and RF = 3000 Ω. R5 takes on the following values: 450 Ω, 900 Ω, and 1350 Ω. What is the quality factor when R5 = 450 Ω? Plot the frequency response.

Solution

PSPICE is used to obtain the frequency response, and MATLAB is employed to obtain the cut-off frequency and to calculate the quality factor.

PSPICE program:

```
*  MODIFIED SALLEN-KEY HIGHPASS FILTER
VIN   1   0   AC     0.5V
VCC   7   0   DC     15V
VEE   8   0   DC     -15V
C1    1   2   0.05e-6
C2    2   3   0.05e-6
R1    2   6   600
R2        3   0       600
R3        4   0       600
R4        6   0       600
```

(continued)

```
RF    5   4   3000
X1 3 4 7 8 5 UA741: UA741 Op Amp
* +INPUT; -INPUT; +VCC; -VEE; OUTPUT; CONNECTIONS FOR UA741
.LIB NOM.LIB;
* UA741 OP AMP MODEL IN PSPICE LIBRARY FILE NOM.LIB
.PARAM   VAL = 900
R5 5 6 {VAL}
.STEP PARAM VAL LIST 450 900 1350
.AC DEC 20 100Hz 100KHz
.PRINT AC VM(5)
.PROBE V(5)
.END
```

Partial results of PSPICE simulation for R5 = 450 Ω are shown in Table 6.6. The complete output can be found in files ex6_6aps.dat, ex6_6bps.dat and ex6_6cps.dat for results for R5 equals to 450 Ω, 900 Ω, and 1350 Ω, respectively.

TABLE 6.6

Output Voltage as a Function
of Frequency for R5 = 450 Ω

Frequency (Hz)	Gain for R5 = 450 Ω (multiply entries by 2)
1.000E+02	3.047E–03
3.162E+02	3.062E–02
5.012E+02	7.751E–02
7.079E+02	1.567E–01
1.000E+03	3.211E–01
3.162E+03	5.876E+00
5.012E+03	1.372E+01
7.079E+03	8.259E+00
1.000E+04	6.655E+00
3.162E+04	5.555E+00
5.012E+04	5.382E+00
7.079E+04	5.187E+00
1.000E+05	4.872E+00

MATLAB script:

```
% load pspice results
load 'ex6_6aps.dat' -ascii;
load 'ex6_6bps.dat' -ascii;
load 'ex6_6cps.dat' -ascii;
```

(continued)

```
fre = ex6_6aps(:,1);
g450 = 2*ex6_6aps(:,2);
g900 = 2*ex6_6bps(:,2);
g1350 = 2*ex6_6cps(:,2);
m = length(fre);
tol = 1.0e-4;
% Plot frequency response
plot(fre,g450, fre,g900, fre, g1350)
xlabel('Frequency, Hz')
ylabel('Gain')
title('Frequency Response of a Sallen-Key Highpass Filter')
%
% Determine quality factor for R% = 450 Ohms
[gmax, kg] = max(g450):  % maximum value of gain
gcf = 0.707 * gmax: % cut-off frequency
% determine the cut-off frequencies
% low cut-off frequency, index, lcf
% High cut-off frequency, index, hcf
k = kg: % initalize counter
while (g450(k) - gcf) > tol
     k = k+ 1;
end
hcf = k;
i = kg;
while (g450(i) - gcf) > tol
     i = i- 1;
end
lcf = i;
% Calculate Quality factor
Qfactor = fre(kg)/(fre(hcf) - fre(lcf))
```

The plot of the frequency response is shown in Figure 6.17. The quality factor equals 2.1528 when R5 is 450 Ω.

6.3.3   Bandpass Filters

A bandpass filter passes a band of frequencies and attenuates other bands. The filter has two cut-off frequencies, f_L and f_H. We assume that $f_H > f_L$. All

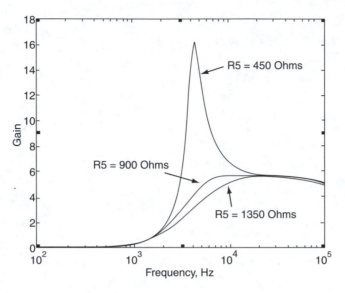

FIGURE 6.17
Frequency response of a modified Sallen-Key highpass filter.

signal frequencies lower than f_L or greater than f_H are attenuated. The general form of the transfer function of a bandpass filter is

$$H(s) = \frac{k\left(\frac{\omega_c}{Q}\right)s}{s^2 + \left(\frac{\omega_c}{Q}\right)s + \omega_c^2}$$

(6.28)

where
    k is the passband gain
    ω_c is the center frequency, in rad/s.

The quality factor Q is related to the 3-dB bandwidth and the center frequency by the expression

$$Q = \frac{\omega_c}{BW} = \frac{f_c}{f_H - f_L}$$

(6.29)

Bandpass filters with $Q \le 10$ are classified as wide bandpass. On the other hand, bandpass filters with $Q > 10$ are considered narrow bandpass.

Wide bandpass filters can be implemented by cascading lowpass and highpass filters. The order of the bandpass filters is the sum of the highpass and lowpass sections. The advantages of this arrangement are that the fall-off, bandwidth, and midband gain can be set independently.

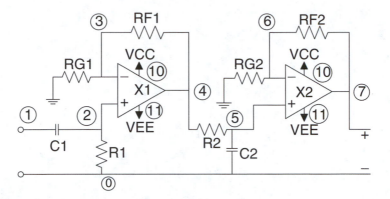

FIGURE 6.18
Second-order wide bandpass filter.

Figure 6.18 shows the wide bandpass filter, built using first-order highpass and first-order lowpass filters. The magnitude of the voltage gain is the product of the voltage gains of both the highpass and lowpass filters.

The following example illustrates the characteristics of the wideband filter.

Example 6.7 Second-Order Wide Bandpass Filter

For the bandpass filter shown in Figure 6.18, assume that op amps X1 and X2 are UA741 op amps. If RG1 = RG2 = 1 KΩ, RF1 = RF2 = 5 KΩ, C1 = 30 nF, R1 = 50 KΩ, R2 = 1000 Ω, and C2 = 15 nF, determine the bandwidth, low cut-off frequency, high cut-off frequency, and quality factor.

Solution

PSPICE is used to obtain the frequency response data.

PSPICE program:

```
BANDPASS FILTER (HIGHPASS PLUS LOWPASS SECTIONS)
VIN      1    0    AC    0.2V
C1       1    2    30E-9
R1       2    0    50E3
RG1      3    0    1E3
RF1      3    4    5E3
X1       2    3    10    11   4   UA741; UA741 OP AMP
* +INPUT; -INPUT; +VCC; -VEE; OUTPUT; CONNECTIONS FOR
UA741
.LIB NOM.LIB;
* UA741 OP AMP MODEL IN PSPICE LIBRARY FILE NOM.LIB
```

(continued)

```
R2          4     5     1.0E3
C2          5     0     15E-9
X2          5     6     10      11    7    UA741; UA741 OP AMP
RG2         6     0     1.0E3
RF2         6     7     5.0E3
VCC         10    0     DC      15V
VEE         11    0     DC      -15V
.AC         DEC   40    10HZ    100KHZ
.PRINT   AC    VM(7)
.PROBE  V(7)
.END
```

Partial results of the PSPICE simulation are shown in Table 6.7. The complete output can be found in file ex6_7ps.dat.

TABLE 6.7

Output Voltage vs. Frequency
for Bandpass Filter

Frequency (Hz)	Gain (multiply entries by 5)
1.000E+01	6.755E–01
5.012E+01	3.075E+00
1.000E+02	4.937E+00
1.000E+02	4.937E+00
5.012E+02	7.035E+00
1.000E+03	7.127E+00
5.012E+03	6.507E+00
1.000E+04	5.230E+00
5.012E+04	1.386E+00
1.000E+05	5.785E–01

The low cut-off frequency, bandwidth, and quality factor are calculated using MATLAB.

MATLAB script:

```
% Load pspice results
load 'ex6_7ps.dat' -ascii;
fre = ex6_7ps(:,1);
vout = 5*ex6_7ps(:,2);
```

(continued)

```
m = length(fre);
% [gmax, kg] = max(g450):  % maximum value of gain
[g_md, m2] = max(vout): % mid-band gain
g_cf = 0.707*g_md:  % gain at cut-off
% Determine the low cut-off frequency index
i = m2;
tol = 1.0e-4;
while (vout(i) - g_cf) > tol;
    i = i - 1;
end
lcf = i;
% Determine the high frequency cut-off index
k = m2;
while (vout(k) - g_cf) > tol;
    k = k + 1;
end
hcf = k;
low_cf = fre(lcf):  % Low cut-off frequency
high_cf = fre(hcf): % High cut-off frequency
ctr_cf = fre(m2):  % Center frequency
band_wd = high_cf - low_cf: % Bandwidth
Qfactor = ctr_cf/band_wd;
% Print results
low_cf
high_cf
band_wd
Qfactor
```

The results are shown below:

Low cut-off frequency is 100 Hz

High cut-off frequency is 11220 Hz

Bandwidth is 11120 Hz

Quality factor is 0.0952

A narrow bandpass filter normally has a high Q-value. A circuit that can be used to implement a narrow band filter is a multiple feedback filter. The circuit is shown in Figure 6.19.

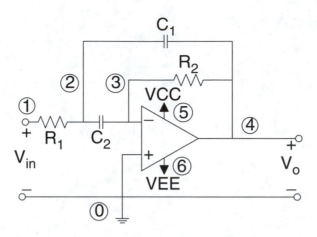

FIGURE 6.19
Multiple-feedback bandpass filter.

The circuit has two feedback paths and one op amp. The circuit can be designed to have a low Q-value and thus manifest wide bandpass filter characteristics. It can be shown that the transfer function of the filter network is

$$
H_{PB} = \frac{V_0(s)}{V_{IN}(s)} = \frac{\left(-\frac{1}{R_1 C_1}\right)s}{s^2 + \left(\frac{1}{R_2}\right)\left(\frac{1}{C_1} + \frac{1}{C_2}\right)s + \frac{1}{R_1 R_2 C_1 C_2}}
\tag{6.30}
$$

$$
= \frac{k_{PB}\left(\omega_C/Q\right)s}{s^2 + \left(\frac{\omega_C}{Q}\right)s + \omega_C^2}
$$

The center frequency f_c is

$$
f_C = \frac{\omega_C}{2\pi} = \frac{1}{2\pi}\frac{1}{\sqrt{R_1 R_2 C_1 C_2}}
\tag{6.31}
$$

If C1 = C2 = C, the Q value is

$$
Q = \frac{1}{2}\sqrt{\frac{R_2}{R_1}}
\tag{6.32}
$$

The passband gain k_{PB} is

$$
k_{PB} = \frac{1}{R_1 C_1}\left(\frac{Q}{\omega_C}\right)
\tag{6.33}
$$

The following example explores the characteristics of the multiple-feedback bandpass filter.

Example 6.8   Multiple-Feedback Narrow Bandpass Filter

For the circuit shown in Figure 6.19, the elements have the following values: R1 = 1 KΩ and C1 = C2 = 100 nF. The op amp is UA741. If R2 takes the values of 25, 75, and 125 KΩ, what are the quality factors, bandwidths, and center frequencies for these values of R2?

Solution

PSPICE is used to obtain the frequency response data, and MATLAB is used for the data analysis.

PSPICE program:

```
NARROWBAND-PASS FILTER (MULTIPLE-FEEDBACK BANDPASS)
.PARAM R2_VAL = 25K;
VIN   1    0    AC        0.1
R1    1    2    1.0E3
C1    2    4    100.0E-9
C2    2    3    100.0E-9
R2    3    4    {R2_VAL}
.STEP PARAM R2_VAL 25K 125K 50K
X1    0    3    5     6    4   UA741
* +INPUT; -INPUT; +VCC; -VEE; OUTPUT; CONNECTIONS FOR UA741
.LIB NOM.LIB;
* UA741 OP AMP MODEL IN PSPICE LIBRARY FILE NOM.LIB
VCC   5    0    DC        15V
VEE   6    0    DC        -15V
.AC   LIN   500   100HZ   1KHZ
.PRINT AC VM(4)
.PROBE V(4)
.END
```

The PSPICE partial results for feedback resistance R2 of 25 KΩ are shown in Table 6.8. The complete PSPICE data can be found in file ex6_8aps.dat, ex6_8bps.dat, and ex6_8cps.dat for R2 equal to 25, 75, and 125 KΩ, respectively.

TABLE 6.8

Frequency Response for Multiple-
Feedback Bandpass Filter for R2 = 25 KΩ

Frequency (Hz)	Gain for R2 = 25KΩ (multiply entries by 1000)
1.000E+02	1.726E–01
1.505E+02	2.959E–01
2.010E+02	4.846E–01
2.515E+02	8.059E–01
3.002E+02	1.201E+00
3.507E+02	1.123E+00
4.012E+02	8.111E–01
4.499E+02	6.148E–01
5.004E+02	4.901E–01
5.509E+02	4.089E–01
6.014E+02	3.520E–01
6.501E+02	3.112E–01

The data analysis is done using MATLAB.

MATLAB script:

```
% Load data
load 'ex6_8aps.dat' -ascii;
load 'ex6_8bps.dat' -ascii;
load 'ex6_8cps.dat' -ascii;
fre = ex6_8aps(:,1);
vo_25K = 1000*ex6_8aps(:,2);
vo_75K = 1000*ex6_8bps(:,2);
vo_125K = 1000*ex6_8cps(:,2);
% Determine center frequency
[vc1, k1] = max(vo_25K)
[vc2, k2] = max(vo_75K)
[vc3, k3] = max(vo_125K)
fc(1) = fre(k1): % center frequency for circuit with R2 = 25K
fc(2) = fre(k2): % center frequency for circuit with R2 = 75K
fc(3) = fre(k3): % center frequency for circuit with R2 =
125K
% Calculate the cut-off frequencies
vgc1 = 0.707 * vc1: % Gain at cut-off for R2 = 25K
vgc2 = 0.707 * vc2: % Gain at cut-off for R2 = 100K
```

(continued)

```
vgc3 = 0.707 * vc3: % Gain at cut-off for R2 = 150K
tol = 1.0e-4:  % tolerance for obtaining cut-off
l1 = k1;
while(vo_25K(l1) - vgc1) > tol
      l1 = l1 +1;
end
fhi(1) = fre(l1):  % high cut-off frequency for R2 = 25K
l1 = k1
while(vo_25K(l1) - vgc1) > tol
      l1 = l1 - 1;
end
flow(1) = fre(l1):  % Low cut-off frequency for R2 = 25K
l2 = k2;
while(vo_75K(l2) - vgc2) > tol
     l2 = l2 + 1;
end
fhi(2) = fre(l2):  % high cut-off frequency for R2 = 75K
l2 = k2;
while(vo_75K(l2) - vgc2) > tol
      l2 = l2 - 1;
end
flow(2) = fre(l2):  % Low cut-off frequency for R2 = 100K
l3 = k3;
while(vo_125K(l3) - vgc3)>tol;
      l3 = l3 + 1;
end
fhi(3) = fre(l3):  % High cut-off frequency
l3 = k3
while(vo_125K(l3) - vgc3) > tol;
      l3 = l3 - 1;
end
flow(3) = fre(l3):  %low cut-off frequency
% Calculate the Quality Factor
for i = 1:3
bw(i) = fhi(i)-flow(i);
Qfactor(i) = fc(i)/bw(i);
```

(continued)

```
end
% Print out results
% Center frequency, high cut-off freq, low cut-off freq
and Q factor are
fc
bw
Qfactor
% plot frequency response
plot(fre,vo_25K, fre,vo_75K, fre, vo_125K)
xlabel('Frequency, Hz')
ylabel('Gain')
title('Frequency Response of a Narrowband Filter')
```

Figure 6.20 shows the frequency response of the narrow bandpass filter.

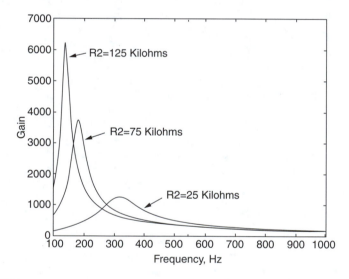

FIGURE 6.20
Narrowband filter response for various values of feedback resistance R2.

The results are shown in Table 6.9.

6.3.4   Band-Reject Filters

A band-reject filter is used to eliminate a specific band of frequencies. It is normally used in communication and biomedical instruments to eliminate unwanted frequencies. The general form of the transfer function of the band-reject filter is

TABLE 6.9

Center Frequency, Bandwidth, and Q factor
for a Narrow Bandpass Filter for Various
Values of Feedback R2

Resistor (KΩ)	Center Frequency (Hz)	Bandwidth (Hz)	Q-factor
25	316.4	128.1	2.47
75	183.0	43.3	4.23
125	141.5	27.0	5.24

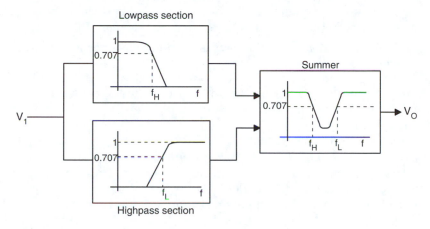

FIGURE 6.21
Block diagram of wideband reject filter.

$$H_{BR} = \frac{k_{PB}\left(s^2 + \omega_C^2\right)}{s^2 + \left(\dfrac{\omega_C}{Q}\right)s + \omega_C^2} \tag{6.34}$$

where
  k_{PB} is the passband gain
  ω_C is the center frequency of the band-reject filter

Band-reject filters are classified as wideband reject ($Q < 10$) and narrow-band reject filters ($Q > 10$). Narrowband reject filters are commonly called notch filters. The wideband reject filter can be implemented by summing the responses of highpass section and lowpass section through a summing amplifier. The block diagram arrangement is shown in Figure 6.21.

The order of the band-reject filter is dependent on the order of lowpass and highpass sections. There are two important requirements for implementing the wideband reject filter using the scheme shown in Figure 6.21:

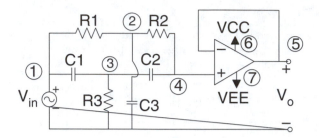

FIGURE 6.22
Narrowband-reject twin-T network.

1. The cut-off frequency f_L of the highpass filter section must be greater than the cut-off frequency f_H of the lowpass filter section.
2. The passband gains of the lowpass and highpass sections must be equal.

A narrowband reject filter or notch filter can be implemented by using a twin-T network. This circuit consists of two parallel T-shaped networks, as shown in Figure 6.22.

Normally, R1 = R2 = R, R3 = R/2, C1 = C2 = C, and C3 = 2C. The R1-C3-R2 section is a lowpass filter with corner frequency $f_c = (4\pi RC)^{-1}$. The C1-R3-C2 is a highpass filter with corner frequency $f_c = (\pi RC)^{-1}$. The center frequency or the notch frequency is $f_c = (2\pi RC)^{-1}$. At the latter frequency, the phases of the two filters cancel out. The following example explores the characteristics of a notch filter.

Example 6.9   Worst-Case Notch Frequency

For the twin-T network shown in Figure 6.22, R1 = R2 = 20 KΩ, R3 = 10 KΩ, C1 = C2 = 0.01 μF, C3 = 0.02 μF, and the op amp is UA741. Determine the nominal notch frequency and the worst-case notch frequency for all devices if tolerances on the resistors and capacitors are 10%. Determine the change in notch frequency due the component tolerances.

Solution

PSPICE is used to perform Monte Carlo analysis. The worst-case notch frequency is obtained from the latter analysis. MATLAB is used to determine the nominal and worst-case notch frequencies. In addition, MATLAB is used to plot the frequency response of the notch filter.

PSPICE program:

```
TWIN-T BANDREJECT FILTER
.OPTIONS RELTOL = 0.10; 10% COMPONENTS
VIN   1   0   AC      1.0V
```

(continued)

```
R1     1   2   RMOD    20K
R2     2   4   RMOD    20K
C3     2   0   CMOD    0.02U
C1     1   3   CMOD    0.01U
C2     3   4   CMOD    0.01U
R3     3   0   RMOD    10K
VCC    6   0   DC      15V
VEE    7   0   DC      -15V
.MODEL RMOD RES(R = 1, DEV = 10%): 10 % RESISTORS
.MODEL CMOD CAP(C = 1, DEV = 10%): 10 % CAPACITORS
X1 4 5 6 7 5 UA741; 741 OP AMP
* +INPUT; -INPUT; +VCC; -VEE; OUTPUT; CONNECTIONS FOR UA741
.LIB NOM.LIB;
* UA741 OP AMP MODEL IN PSPICE LIBRARY FILE NOM.LIB
.AC DEC 100 10HZ 100KHZ
.WCASE AC V(5) MAX OUTPUT ALL; SENSITIVITY & WORST CASE
ANALYSIS
.PRINT AC VM(5)
.PROBE V(5)
.END
```

PSPICE partial results for nominal values of the capacitors and resistors are shown in Table 6.10. The complete PSPICE output data can be found in files ex6_9aps.dat and ex6_9bps.dat for nominal and worst-case for all devices, respectively.

TABLE 6.10

Frequency Response of Twin-T
Notch Filter for Nominal Values
of Resistors and Capacitors

Frequency (Hz)	Output Voltage for 10% Device Tolerance (V)
1.000E+01	9.987E–01
5.012E+01	9.694E–01
1.000E+02	8.905E–01
5.012E+02	2.329E–01
1.000E+03	1.145E–01
5.012E+03	8.378E–01
1.000E+04	9.523E–01
5.012E+04	9.981E–01
1.000E+05	1.000E+00

MATLAB is used for data analysis.

MATLAB script:

```
% Load the data
load 'ex6_9aps.dat' -ascii;
load 'ex6_9bps.dat' -ascii;
fre = ex6_9aps(:,1);
vo_nom = ex6_9aps(:,2);
vo_wc = ex6_9bps(:,2);
%
% Determination of center frequency
[vc(1), k(1)] = min(vo_nom);
[vc(2), k(2)] = min(vo_wc);
for i = 1:2
fc(i) = fre(k(i));
end
% Determine difference between center frequencies
fc_dif = fc(1) - fc(2);
% Plot the frequency response
plot(fre, vo_nom, fre, vo_wc);
xlabel('Frequency, Hz')
ylabel('Gain')
title('Frequency Response of a Notch Filter')
fc
fc_dif
```

The frequency response of the notch filter is shown in Figure 6.23. The results from MATLAB are

The center frequency of the filter with nominal component values is 794.3 Hz

The center frequency of the filter with all worst-case component values is 707.9 Hz

The difference between the above two center frequencies is 86.4 Hz

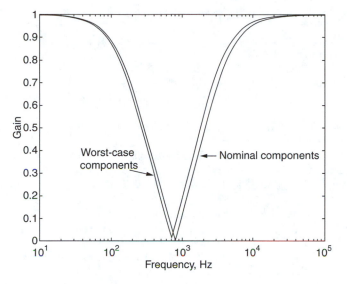

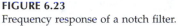

FIGURE 6.23
Frequency response of a notch filter.

Bibliography

1. Al-Hashimi, Bashir, *The Art of Simulation Using PSPICE, Analog, and Digital,* CRC Press, Boca Raton, FL, 1994.
2. Attia, J.O., *Electronics and Circuit Analysis Using MATLAB,* CRC Press, Boca Raton, FL, 1999.
3. Biran, A. and Breiner, M., *MATLAB for Engineers,* Addison-Wesley, Reading, MA, 1995.
4. Boyd, Robert R., Tolerance Analysis of Electronic Circuits Using MATLAB, CRC Press, Boca Raton, FL, 1999.
5. Chapman, S.J., MATLAB Programming for Engineers, Brook, Cole Thompson Learning, Pacific Grove, CA, 2000.
6. Derenzo, S.E., Interfacing: A Laboratory Approach Using the Microcomputer for Instrumentation, Data Analysis and Control, Prentice-Hall, Englewood Cliffs, NJ, 1990.
7. Distler, R.J., Monte Carlo Analysis of System Tolerance, *IEEE Trans. Education,* 20, 98–101, May 1997.
8. Ellis, George, Use SPICE to Analyze Component Variations in Circuit Design, *EDN,* pp. 109–114, April 1993,.
9. Etter, D.M., *Engineering Problem Solving with MATLAB,* 2nd edition, Prentice-Hall, Upper Saddle River, NJ, 1997.
10. Etter, D.M., Kuncicky, D.C., and Hull, D., *Introduction to MATLAB 6,* Prentice-Hall, Upper Saddle River, NJ, 2002.
11. Fenical, L.H., *PSPICE: A Tutorial,* Prentice-Hall, Englewood Cliffs, NJ, 1992.

12. Gottling, J.G., *Matrix Analysis of Circuits Using MATLAB,* Prentice-Hall, Englewood Cliffs, NJ, 1995.
13. Hamann, J.C., Pierre, J.W., Legowski, S.F., and Long, F.M., Using Monte Carlo Simulations to Introduce Tolerance Design to Undergraduates, *IEEE Trans. Education,* 42(1), 1–14, Feb. 1999.
14. Kavanaugh, Micheal F., Including the Effects of Component Tolerances in the Teaching of Courses in Introductory Circuit Design, *IEEE Trans. Education,* 38(4), 361–364, Nov. 1995.
15. Keown, John, *PSPICE and Circuit Analysis,* Maxwell MacMillan International Publishing Group, New York, 1991.
16. Kielkowski, Ron M., *Inside SPICE, Overcoming the Obstacles of Circuit Simulation,* McGraw-Hill, New York, 1994.
17. Lamey, Robert, *The Illustrated Guide to PSPICE,* Delmar Publishers, Albany, NY, 1995.
18. Nilsson, James W. and Riedel, Susan A., *Introduction to PSPICE,* Addison-Wesley, Reading, MA, 1993.
19. OrCAD PSPICE A/D, Users' Guide, November 1998.
20. Prigozy, Stephen, Novel Applications of PSPICE in Engineering, *IEEE Trans. Education,* 32(1), 35–38, Feb. 1989.
21. Rashid, Mohammad H., *Microelectronic Circuits, Analysis and Design,* PWS Publishing Company, Boston, MA, 1999.
22. Rashid, Mohammad H., *SPICE for Circuits and Electronics Using PSPICE,* Prentice-Hall, Englewood Cliffs, NJ, 1990.
23. Roberts, Gordon W. and Sedra, Adel S., *SPICE for Microelectronic Circuits,* Saunders College Publishing, Fort Worth, TX, 1992.
24. Sedra, A.S. and Smith, K.C., *Microelectronic Circuits,* 4th edition, Oxford University Press, 1998.
25. Sigmor, K., *MATLAB Primer,* 4th edition, CRC Press, Boca Raton, FL, 1998.
26. Spence, Robert and Soin, Randeep S., *Tolerance Design of Electronic Circuits,* Imperial College Press, River Edge, NY, 1997.
27. Thorpe, Thomas W., *Computerized Circuit Analysis with SPICE,* John Wiley & Sons, New York, 1991.
28. Tuinenga, Paul W., *SPICE, A Guide to Circuit Simulations and Analysis Using PSPICE,* Prentice-Hall, Englewood Cliffs, NJ, 1995.
29. Using MATLAB, The Language of Technical Computing, Computation, Visualization, and Programming, Version 6, MathWorks, Inc. 2000.
30. Vladimirescu, Andrei, *The SPICE Book,* John Wiley & Sons, New York, 1994.

Problems

6.1 For the inverting amplifier configuration shown in Figure P6.1, R1 = 1 KΩ, VCC = 15 V, and VEE = –15 V. R2 takes values of 5, 10, 20, 30, 40, and 50 KΩ. Assume that the op amp is UA741.

   (a) Determine the corresponding low-frequency gain and the cut-off frequency.
   (b) Plot a graph of cut-off frequency vs. gain.
   (c) What is the unity gain bandwidth?

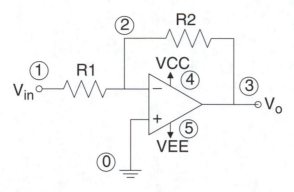

FIGURE P6.1
Inverting amplifier.

6.2 For Figure 6.5, X1 is a UA741 op amp, VCC = 15 V, VEE = –15 V, R1 = 1 KΩ, and R2 = 4 KΩ. Determine the 3-dB frequency and the unity-gain bandwidth.

6.3 A four-pole Sallen-Key lowpass filter is shown in Figure P6.3. VCC = 15 V, VEE = –15 V, R = 3 KΩ, R1 = R3 = 10 KΩ, R2 = 150 Ω, R4 = 12 KΩ, and C = 0.01 μF. Assume UA741 op amps.

(a) Find the cut-off frequency.

(b) What is the gain in the passband?

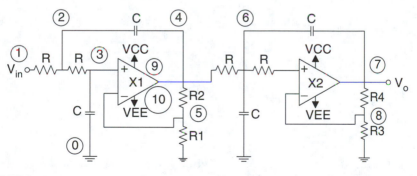

FIGURE P6.3
Sallen-Key lowpass filter.

6.4 The modified Sallen-Key circuit shown in Figure P6.4 is similar to the Sallen-Key circuit shown in Figure 6.12, except that a voltage divider network has been added at the output. This addition results in a higher dc gain. In addition, the cut-off frequency is affected. This exercise explores the gain as a function of the cut-off frequency. Assume that R1 = R2 = 30 KΩ, R3 = 10 KΩ, R4 = 40 KΩ, C1 = C2 = 0.05 μF, R5 = 10 KΩ, VCC = 15 V, and VEE = –15 V. Op amp is UA741. Determine the cut-off frequency and the maximum gain when R6 takes the following values: 10, 8, 6, and 4 KΩ.

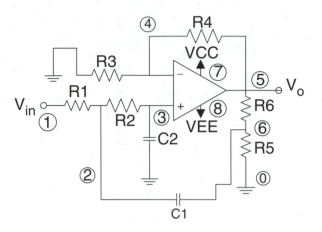

FIGURE P6.4
Modified Sallen-Key lowpass filter.

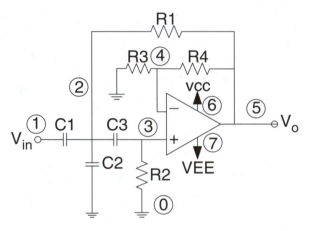

FIGURE P6.5
Butterworth highpass filter.

6.5 A Butterworth second-order highpass filter is shown in Figure P6.5. If $C3 = 0.05 \ \mu F$, $R1 = R2 = R3 = 600 \ \Omega$, $R4 = 400 \ \Omega$, $C1 = 30$ nF, and $C2 = 19$ nF, determine the cut-off frequency and the gain at the cut-off frequency. Assume a UA741 op amp.

6.6 In Example 6.6, if R5 has the values $450 \ \Omega$, $470 \ \Omega$, $490 \ \Omega$, $510 \ \Omega$, and $530 \ \Omega$, determine the quality factor for each value of R5. Plot the quality factor as a function of R5. Assume UA741 op amp.

6.7 In Figure P6.7, $RG1 = RG2 = 1 \ K\Omega$, $RF1 = RF2 = 3 \ K\Omega$, $R1 = R2 = 45 \ K\Omega$, $R3 = R4 = 100 \ \Omega$, $C1 = C2 = 25$ nF, and $C3 = C4 = 10$ nF. Assume that X1 and X2 are 741 op amps. Calculate the high-frequency cut-off. Determine the low-frequency cut-off. Calculate the bandwidth. What is gain in the midband of the filter?

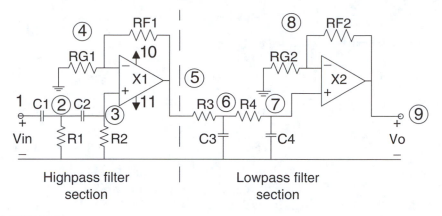

Highpass filter
section

Lowpass filter
section

FIGURE P6.7
Wide bandpass filter implemented by cascading highpass and lowpass filters.

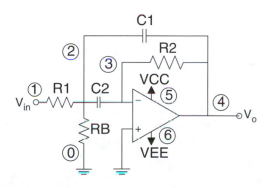

FIGURE P6.8
Modified multiple-feedback bandpass filter.

6.8 Figure P6.8 is a modified multiple-feedback bandpass filter similar to Figure 6.19. R1 = 10 KΩ, C1 = C2 = 1 nF, and R2 = 150 KΩ. If RB is 5 KΩ, 10 KΩ, and 15 KΩ, find the Q-value, center frequency and bandwidth for each value of RB. Assume that VCC = 15 V, VEE = –15 V, and the op amp is UA741.

6.9 Figure P6.9 is a wideband reject filter obtained using lowpass and highpass sections. VCC = 15 V, VEE = –15 V, R3 = R5 = 1 KΩ, R4 = R6 = 5 KΩ, R7 = R8 = R9 = 2 KΩ, R1 = 100 Ω, R2 = 100 KΩ, and C1 = C2 = 1 μF. Determine the notch frequency, bandwidth, and quality factor. Assume that X1, X2, and X3 are UA741 op amps.

6.10 For the twin-T network shown in Figure 6.22, R1 = R2 = 10 KΩ, R3 = 5 KΩ, C1 = C2 = 0.01 μF, C3 = 0.02 μF, and the op amp is UA741. Determine the notch frequency, the bandwidth of the filter, the quality factor, and worst-case notch frequency if the resistors and capacitors have tolerances of 5%.

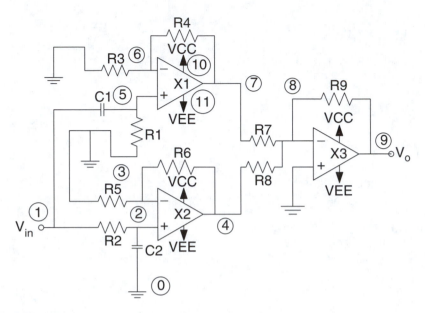

FIGURE P6.9
Wideband band-reject filter.

6.11 In Example 6.9, the component tolerance is changed to 15%. Find the notch frequency for the nominal values of the capacitors and resistors. Determine the worst-case notch frequency. For the nominal component values, determine the bandwidth and quality factor of the notch filter.

7

Transistor Characteristics and Circuits

This chapter discusses the characteristics of bipolar junction transistors and metal oxide semiconductor field effect transistors. The sensitivity of the transistor biasing circuits is explored through PSPICE simulations and MATLAB calculations. The frequency response of amplifiers and feedback amplifiers are also discussed.

7.1 Characteristics of Bipolar Junction Transistors

A bipolar junction transistor (BJT) consists of two pn junctions connected back-to-back. The operation of the BJT depends on the flow of both majority and minority carriers. The dc behavior of the BJT can be described by the Ebers-Moll model. The voltages of the base-emitter and base-collector junctions define the region of operation of the BJT. The regions of normal operation are forward-active, reverse-active, saturation, and cut-off. Table 7.1 shows the regions of operation based on the polarities of the base-emitter and base-collector junctions.

The forward-active region corresponds to forward biasing the emitter-base junction and reverse biasing the base-collector junction. It is the normal operation region of bipolar junction transistors when employed for amplifications. In the forward-active region, the first-order representations of the collector current I_C and base current I_B are given as

$$I_C = I_S \exp\left(\frac{V_{BE}}{V_T}\right)\left(1 + \frac{V_{CE}}{V_{AF}}\right) \tag{7.1}$$

and

$$I_B = \frac{I_S}{\beta_F} \exp\left(\frac{V_{BE}}{V_T}\right) \tag{7.2}$$

TABLE 7.1

Regions of Operation of BJT

Base-Emitter Junction	Base-Collector Junction	Region of Operation
Forward bias	Reverse bias	Forward active
Forward bias	Forward bias	Saturation
Reverse bias	Reverse bias	Cut-off

where

β_F is large signal forward current gain of common-emitter configuration
V_{AF} is forward early voltage
I_S is the BJT transport saturation current
V_T is the thermal voltage given by

$$V_T = \frac{kT}{q} \tag{7.3}$$

where

k is the Boltzmann constant ($k = 1.381 \times 10^{-23}$ VC/°K).
T is the absolute temperature in degrees Kelvin
q is the charge of an electron ($q = 1.602 \times 10^{-19}$ C)

If $V_{AF} \gg V_{CE}$, then from Equations (7.1) and (7.2), we have

$$I_C = \beta_F I_B \tag{7.4}$$

The saturation region corresponds to forward biasing both base-emitter and base-collector junctions. The cut-off region corresponds to reverse biasing the base-emitter and base-collector junctions. In the cut-off region, the collector and base currents are insignificant compared to those that flow when transistors are in the active-forward and saturation regions.

From Equation (7.2), the input characteristic of a forward biased base-emitter junction is similar to diode characteristics. In the following example, we explore the output characteristics of a BJT transistor.

Example 7.1   BJT Output Characteristics

Figure 7.1 shows an arrangement that can be used to find the output characteristics of transistor Q2N2222. Draw I_C vs. V_{CE} for $I_B = 2$ μA, 4 μA, and 6 μA. Calculate the output resistance (r_{CE}) as a function of V_{CE} for $I_B = 2$ μA. Assume that R1 = R2 = R3 = 1 Ω.

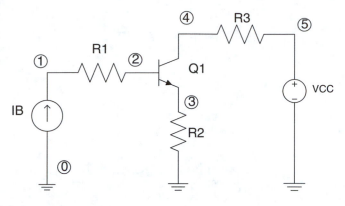

FIGURE 7.1
Circuit for obtaining BJT output characteristics.

Solution

PSPICE is used to obtain the data needed for plotting the current vs. voltage characteristics of transistor Q2N2222. MATLAB is used to plot the output characteristics and also to calculate the output resistance.

PSPICE program:

```
BJT CHARACTERISTICS
VCC   5  0   DC   0V
R1    1  2   1
R2       3  0   1
R3       5  4   1
IB   0  1   DC   6UA
Q1       4  2   3      Q2N2222
.MODEL Q2N2222 NPN(BF=100 IS=3.295E-14 VA=200); TRANSISTOR
MODEL
** ANALYSIS TO BE DONE
** VARY VCE FROM 0 TO 10V IN STEPS 0.1V
** VARY IB FROM 2 TO 6mA IN STEPS OF 2mA
.DC VCC 0V 10V .05V IB 2U 6U 2U
.PRINT DC V(4,3) I(R1) I(R3)
.PROBE V(4,3) I(R3)
.END
```

PSPICE partial results for a base current of 2 μA are shown in Table 7.2. The complete results can be found in files ex7_1aps.dat, ex7_1bps.dat, and ex7_1cps.dat for base currents of 2, 4, and 6 μA, respectively.

TABLE 7.2

Output Characteristics
of Transistor Q2N2222

V_{CE} (V)	I_C (A)
4.996E–01	1.999E–04
1.050E+00	2.005E–04
2.000E+00	2.014E–04
3.000E+00	2.024E–04
4.000E+00	2.034E–04
5.000E+00	2.044E–04
6.000E+00	2.054E–04
7.000E+00	2.064E–04
8.000E+00	2.074E–04
9.000E+00	2.084E–04
1.000E+01	2.094E–04

MATLAB is used to plot the output characteristics.

MATLAB script:

```
% Load data
load 'ex7_1aps.dat' -ascii;
load 'ex7_1bps.dat' -ascii;
load 'ex7_1cps.dat' -ascii;
vce1 = ex7_1aps(:,2);
ic1 = ex7_1aps(:,4);
vce2 = ex7_1bps(:,2);
ic2 = ex7_1bps(:,4);
vce3 = ex7_1cps(:,2);
ic3 = ex7_1cps(:,4);
plot(vce1, ic1, vce2, ic2, vce3, ic3)
xlabel('Collector-emitter Voltage, V')
ylabel('Collector Current, A')
title('Output Characteristics')
```

The output characteristics are shown in Figure 7.2.

The output resistance r_{CE} as a function of V_{CE} is obtained using MATLAB.

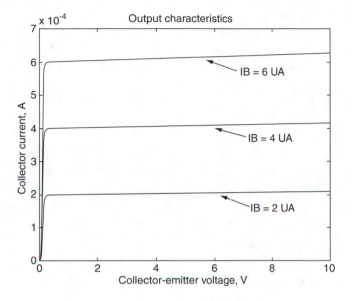

FIGURE 7.2
I_C vs. V_{CE} of transistor Q2N2222.

MATLAB script:

```
% Load data
load 'ex7_1aps.dat' -ascii;
vce = ex7_1aps(:,2);
ic = ex7_1aps(:,4);
m = length(vce):   % size of vector vce
% calculate output resistance
for i = 2:m-1
    rce(i) = (vce(i+1)-vce(i-1))/(ic(i+1)- ic(i - 1)):
% output resistance
end
rce(1) = rce(2);
rce(m) = rce(m-1);
plot(vce(2:m-1), rce(2:m-1))
xlabel('Collector-emitter Voltage')
ylabel('Output Resistance')
title('Output Resistance as a function of Collector-
emitter Voltage')
```

The plot of the output resistance is shown in Figure 7.3.

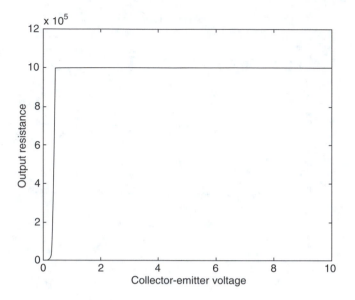

FIGURE 7.3
Output resistance as a function of collector-emitter voltage for Q2N2222 when base current is 2 μA.

7.2   MOSFET Characteristics

Metal-oxide semiconductor field effect transistors (MOSFETs) normally have high input resistance because of the oxide insulation between the gate and the channel. There are two types of MOSFETs: the enhancement type and the depletion type. In the enhancement type, the channel between the source and drain has to be induced by applying a voltage at the gate. In the depletion-type MOSFET, the structure of the device is such that there exists a channel between the source and drain. Because the enhancement-type MOSFET is widely used, the presentation of this section will be done using the enhancement-type MOSFET.

The voltage needed to create the channel between the source and drain is called the threshold voltage V_T. For an n-channel enhancement MOSFET, V_T is positive and for a p-channel device, it is negative.

MOSFETs can operate in three modes: cut-off, triode, and saturation regions. The following is a short description of the three regions of operation.

Cut-off Region:
For an n-channel MOSFET, if the gate-source voltage V_{GS} satisfies the condition

$$V_{GS} < V_T \tag{7.5}$$

then the device is cut off. This implies that the drain current is zero for all values of the drain-to-source voltage.

Triode Region:
When $V_{GS} \geq V_T$ and V_{DS} is small, the MOSFET will be in the triode region. In this region, the device behaves as nonlinear voltage-controlled resistance. The drain current I_D is related to drain-source voltage V_{DS} by

$$I_D = k_n \left[2(V_{GS} - V_T)V_{DS} - V_{DS}^2 \right](1 + \lambda V_{DS}) \tag{7.6}$$

provided that

$$V_{DS} \leq V_{GS} - V_T \tag{7.7}$$

where

$$k_n = \frac{\mu_n \varepsilon \varepsilon_{OX}}{2t_{OX}} \frac{W}{L} = \frac{\mu_n C_{OX}}{2} \left(\frac{W}{L} \right) \tag{7.8}$$

and
   μ_n is the surface mobility of electrons
   ε is the permittivity of free space ($8.85 \times 10^{-12}\,\text{F/cm}$)
   ε_{OX} is the dielectric constant of SiO_2
   t_{OX} is the oxide thickness
   L is the length of the channel
   W is the width of the channel
   λ is the channel width modulation factor

Saturation Region:
If $V_{GS} > V_T$, a MOSFET will operate in the saturation region provided

$$V_{DS} \geq V_{GS} - V_T \tag{7.9}$$

In the saturation region, the current-voltage characteristics are given as

$$I_D = k_n (V_{GS} - V_T)^2 (1 + \lambda V_{DS}) \tag{7.10}$$

The transconductance is given as:

$$g_m = \frac{\Delta I_D}{\Delta V_{GS}} \tag{7.11}$$

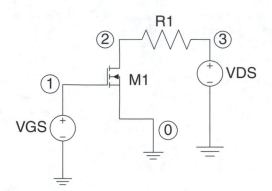

FIGURE 7.4
Circuit for obtaining MOSFET characteristics.

and the incremented drain-to-source resistance r_{DS} is given as

$$r_{DS} = \frac{\Delta V_{DS}}{\Delta I_{DS}} \tag{7.12}$$

The following example obtains the I_D vs. V_{GS} characteristics of a MOSFET.

Example 7.2 Current vs. Voltage Characteristics of a MOSFET

Figure 7.4 shows an arrangement for obtaining the I_D vs. V_{GS} characteristics of a MOSFET. Draw I_D vs. V_{GS} curve, and obtain the transconductance vs. V_{GS}. Assume that M1 is M2N4351.

Solution

PSPICE simulation is used to obtain the corresponding current vs. voltage values of the MOSFET.

PSPICE program:

```
*  ID VERSUS VGS CHARACTERISTICS OF A MOSFET
VDS 3   0   DC 5V
R1   3   2   1
VGS 1   0   DC 2V: THIS IS AN ARBITRARY VALUE, VGS WILL
BE SWEPT
M1   2   1   0   0   M2N4531: NMOS MODEL
.MODEL M2N4531 NMOS(KP=125U VTO=2.24 L=10U W=59U
LAMBDA=5M)
.DC VGS 0V 5V 0.05V
```

(continued)

```
**  OUTPUT COMMANDS
.PRINT DC I(R1)
.PROBE V(1) I(R1)
.END
```

Partial results of the PSPICE simulation are shown in Table 7.3. The complete results can be found in file ex7_2ps.dat.

TABLE 7.3

D_D vs. V_{GS} of MOSFET M2N4351

V_{GS} (V)	I_D (A)
0.000E+00	5.010E–12
5.000E–01	5.010E–12
1.000E+00	5.010E–12
1.500E+00	5.010E–12
2.000E+00	5.010E–12
2.500E+00	2.555E–05
3.000E+00	2.183E–04
3.500E+00	6.001E–04
4.000E+00	1.171E–03
4.500E+00	1.931E–03
5.000E+00	2.879E–03

MATLAB is used to plot I_D vs. V_{GS} of the MOSFET. In addition, the transconductance g_m vs. V_{GS} is obtained using MATLAB.

MATLAB script:

```
% Load pspice data
load 'ex7_2ps.dat' -ascii;
vgs = ex7_2ps(:,1);
ids = ex7_2ps(:,2);
m = length(vgs): % size of vector vgs
% Plot Ids vs. VGS
subplot(211)
plot(vgs, ids)
ylabel('Drain Current, A')
title('Input Characteristics of a MOSFET ')
% Calculate transconductance
for i = 2:m - 1;
```

(continued)

```
      gm(i) = (ids(i+1) - ids(i-1))/(vgs(i+1) - vgs(i-1)):
% transconductance
end
gm(1) = gm(2);
gm(m) = gm(m - 1);
% Plot transconductance
subplot(212);
plot(vgs(2:m - 1), gm(2:m - 1))
xlabel('Gate-to-Source Voltage, V')
ylabel('Transconductanc, A/V')
title('Transconductance vs. Gate-source Voltage')
```

The plots are shown in Figure 7.5.

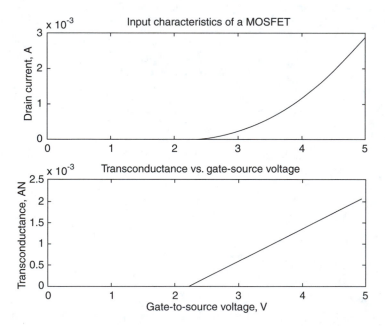

FIGURE 7.5
(Upper) I_D vs. V_{GS}, and (Lower) g_m vs. V_{GS} of MOSFET M2N4351.

7.3   Biasing of BJT Circuits

Biasing networks are used to establish an appropriate dc operating point for transistors in a circuit. For stable and consistent operation, the dc operating point should be held relatively constant under varying conditions. There are

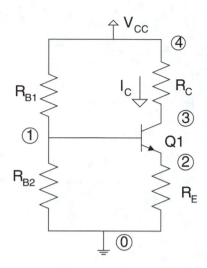

FIGURE 7.6
Biasing circuit for BJT discrete circuits with two base resistors.

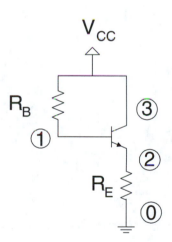

FIGURE 7.7
Biasing BJT discrete network with one base resistor.

several biasing circuits available in the literature. Some are for biasing discrete circuits and others for integrated circuits. Figures 7.6 and 7.7 show some biasing networks for discrete circuits.

Biasing networks for discrete circuits are not suitable for integrated circuits (ICs) because of the large number of resistors and large coupling and bypass capacitors required for biasing discrete electronic circuits. It is uneconomical to fabricate large IC resistors because they take a disproportionately large area on an IC chip. For ICs, biasing is done using mostly transistors that are connected to create constant current sources. Some biasing circuits for integrated circuits are shown in Figures 7.8, 7.9, and 7.10.

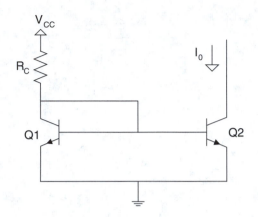

FIGURE 7.8
Simple current mirror for IC biasing.

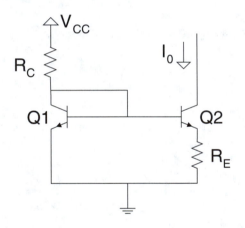

FIGURE 7.9
Widlar current source.

For the bias network for discrete circuits shown in Figure 7.6, it can be shown that

$$I_C = \frac{V_{BB} - V_{BE}}{\dfrac{R_B}{\beta_F} + \dfrac{(\beta_F + 1)}{\beta_F} R_E} \qquad (7.13)$$

and

$$V_{CE} = V_{CC} - I_C \left(R_C + \frac{R_E}{\alpha_F} \right) \qquad (7.14)$$

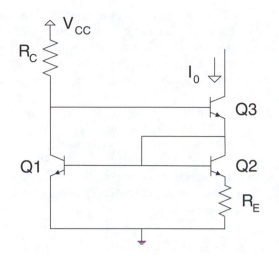

FIGURE 7.10
Wilson current source.

where

$$V_{BB} = \frac{V_{CC}R_{B2}}{R_{B1} + R_{B2}} \qquad (7.15)$$

$$R_B = R_{B1} // R_{B2} = \frac{R_{B1}R_{B2}}{R_{B1} + R_{B2}} \qquad (7.16)$$

$$\alpha_F = \frac{\beta_F}{\beta_F + 1} \qquad (7.17)$$

β_F is a large signal forward current gain of common-emitter configuration.
  For the simple current mirror circuit shown in Figure 7.8, it can also be shown that

$$I_0 = \frac{\beta_F}{\beta_F + 2} I_R \qquad (7.18)$$

where

$$I_R = \frac{V_{CC} - V_{BE}}{R_C} \qquad (7.19)$$

Equation (7.13) gives the parameters that influence the bias current, I_C. Using a stabilized voltage supply, we can ignore changes in V_{CC}, and hence V_{BB}. The changes in resistances R_B and R_E are negligible. However, there is a variation of β_F with respect to changes in I_C. In addition, there is variation of β_F of a specified transistor when selected from different lots or fabricated by different manufacturers.

DC stability of circuit when components vary can be investigated through a dc sensitivity analysis. The PSPICE command for performing sensitivity analysis is the **.SENS** statement, which was discussed in Chapter 1. As mentioned in Chapter 1, the **.SENS** statement allows PSPICE to compute the derivatives of preselected variables of the circuit to most of the components in the circuit. The following example explores the sensitivity of the bias point to components of a biasing network.

Example 7.3   Sensitivity of Collector Current to Amplifier Components

For the common-emitter biasing network shown in Figure 7.6, $V_{CC} = 10$ V, $R_{B1} = 40$ KΩ, $R_{B2} = 10$ KΩ, $R_E = 1$ KΩ, $R_C = 6$ KΩ, and Q1 is Q2N2222. Find the sensitivity of the collector current to amplifier components. Two other Q2N2222 transistors are picked from a lot and they have β_F of 125 and 150. What is the change in I_C with respect to β_F? Use the following model for transistor Q2N2222:

.MODEL   Q2N2222   NPN(BF = 100 IS = 3.295E–14 VA = 200)

Solution

The bias sensitivity is obtained for Q2N2222 when $\beta_F = 100$.

PSPICE program:

```
* SENSITIVITY OF COLLECTOR CURRENT TO AMPLIFIER COMPONENT
VCC  4   0   DC 10V
RB1  4   1   40K
RB2  1   0   10K
RE   2   0   1K
RC   5   3   6K
VM   4   5   DC 0: MONITOR COLLECTOR CURRENT
Q1   3   1   2   Q2N2222
.MODEL Q2N2222 NPN(BF=100 IS=3.295E-14 VA=200)
* ANALYSIS TO BE DONE
.SENS I(VM)
.END
```

The following edited results are obtained from the PSPICE simulation.

```
VOLTAGE SOURCE CURRENTS
        NAME     CURRENT
        VCC      -1.460E-03
        VM        1.257E-03

DC SENSITIVITIES OF OUTPUT I(VM)
        ELEMENT      ELEMENT       ELEMENT        NORMALIZED
        NAME         VALUE         SENSITIVITY    SENSITIVITY
                                   (AMPS/UNIT)    (AMPS/PERCENT)

        RB1     4.000E+04    -3.632E-08      -1.453E-05
        RB2     1.000E+04     1.363E-07       1.363E-05
        RE      1.000E+03    -1.139E-06      -1.139E-05
        RC      6.000E+03    -7.796E-10      -4.678E-08
        VCC     1.000E+01     1.800E-04       1.800E-05
        VM      0.000E+00    -6.202E-07       0.000E+00
Q1
        RB      0.000E+00     0.000E+00       0.000E+00
        RC      0.000E+00     0.000E+00       0.000E+00
        RE      0.000E+00     0.000E+00       0.000E+00
        BF      1.000E+02     1.012E-06       1.012E-06
        ISE     0.000E+00     0.000E+00       0.000E+00
        BR      1.000E+00    -2.692E-13      -2.692E-15
        ISC     0.000E+00     0.000E+00       0.000E+00
        IS      3.295E-14     7.044E+08       2.321E-07
        NE      1.500E+00     0.000E+00       0.000E+00
        NC      2.000E+00     0.000E+00       0.000E+00
        IKF     0.000E+00     0.000E+00       0.000E+00
        IKR     0.000E+00     0.000E+00       0.000E+00
        VAF     2.000E+02    -1.730E-09      -3.460E-09
        VAR     0.000E+00     0.000E+00       0.000E+00
```

The above simulation is performed with two values of β_F: 125 and 150. Table 7.4 shows I_C vs. β_F.

It can be seen from Table 7.4 that as β_F increases, I_C increases.

TABLE 7.4

I_C vs. β_F

β_F	I_C (mA)
100	1.257
125	1.278
150	1.292

Temperature Effects

Temperature changes cause two transistor parameters to change: the base-emitter voltage (V_{BE}) and the collector leakage current between the base and collector (I_{CBO}). For silicon transistors, the voltage V_{BE} varies almost linearly with temperature as:

$$\Delta V_{BE} \cong -2(T_2 - T_1)mV \tag{7.20}$$

where T_1 and T_2 are temperatures in degrees Celsius.

The collector-to-base leakage current I_{CBO} approximately doubles every 10°C temperature rise. From Equations (7.13), (7.18), and (7.19), both I_C and I_O are dependent on V_{BE}. Thus, the bias currents are temperature dependent. The following example explores the sensitivity of the collector current to temperature variation.

Example 7.4   Sensitivity of Common-Collector Amplifier to Temperature

In Figure 7.7, $R_B = 40\ K\Omega$, $R_E = 2\ K\Omega$, Q1 is Q2N3904, and $V_{CC} = 10$ V. Assume a linear dependence of resistance R_B and R_E on temperature and TC1 = 1000 ppm/°C. Determine the emitter current as a function of temperature (0°C to 100°C).

Solution

The PSPICE .DC TEMP command is used to sweep temperature from 0°C to 100°C in steps of 10°C. The changes in resistance due to temperature are calculated by SPICE using the equation:

$$R[T_2] = R(T_1)\left[1 + TC1(T_2 - T_1) + TC2(T_2 - T_1)^2\right] \tag{7.21}$$

where
 $T_1 = 27°C$
 T_2 is the required temperature

TC1 and TC2 are included in the model statement of resistors.

PSPICE program:

```
EMITTER CURRENT DEPENDENCE ON TEMPERATURE
VCC 3  0  DC 10V
RB  3  1  RMOD3 40K: RB IS MODELED
RE  2  0  RMOD3 2K: RE IS MODELED
.MODEL RMOD3 RES(R=1 TC1=1000U TC2=0): TEMP MODEL OF
RESISTORS
Q1 3 1 2 Q2N3904: TRANSISTOR CONNECTIONS
.MODEL Q2N3904 NPN(IS=1.05E-15 ISE=4.12N NE=4 ISC=4.12N
NC=4 BF=220
+ IKF=2E-1 VAF=80 CJC=4.32P CJE=5.27P RB=5 RE=0.5 RC=1
TF=0.617N
+ TR=200N KF=1E-15 AF=1)
* ANALYSIS TO BE DONE
.DC TEMP 0 100 5: VARY TEMP FROM 0 TO 100 IN STEPS OF 5
.PRINT DC I(RE)
.END
```

Table 7.5 shows PSPICE partial results. The complete data can be found in file ex7_4ps.dat.

TABLE 7.5

Temperature vs. Emitter Current

Temperature (°C)	Emitter Current (A)
0.000E+00	4.230E–03
1.000E+01	4.193E–03
2.000E+01	4.156E–03
3.000E+01	4.121E–03
4.000E+01	4.086E–03
5.000E+01	4.053E–03
6.000E+01	4.019E–03
7.000E+01	3.987E–03
8.000E+01	3.955E–03
9.000E+01	3.924E–03
1.000E+02	3.894E–03

MATLAB is used to plot the PSPICE results.

MATLAB script:

```
% Load data
load 'ex7_4ps.dat' -ascii;
temp = ex7_4ps(:,1);
ie = ex7_4ps(:,2);
% plot ie, vs. temp
plot(temp, ie, temp, ie, 'ob')
xlabel('Temperature, °C')
ylabel('Emitter Current, A')
title('Variation of Emitter Current with Temperature')
```

The plot is shown in Figure 7.11.

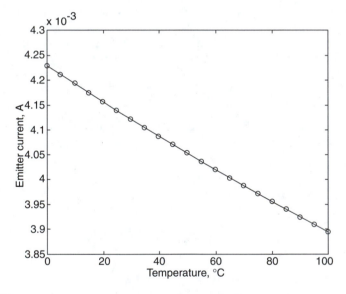

FIGURE 7.11
Emitter current as a function of temperature.

7.4   MOSFET Bias Circuit

There are several circuits that can be used to bias MOSFETs at a stable operating point such that the bias point does not change significantly with changes in MOSFET parameters. Some biasing circuits for MOSFET discrete circuits are shown in Figures 7.12, 7.13, and 7.14.

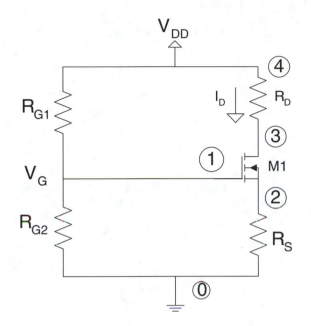

FIGURE 7.12
Biasing circuit for MOSFET using fixed gate voltage and self-bias resistors R_S.

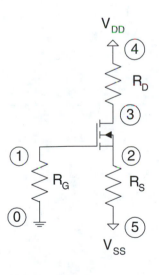

FIGURE 7.13
MOSFET biasing circuit using two power supplies.

In Figure 7.12, it can be shown that

$$V_{GS} = \frac{R_{G1}}{R_{G1} + R_{G2}} V_{DD} - I_D R_S \qquad (7.22)$$

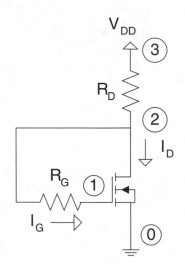

FIGURE 7.14
Biasing MOSFET circuit with resistance feedback using resistor R_G.

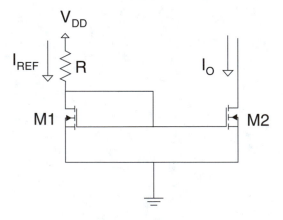

FIGURE 7.15
Basic MOSFET current source.

and

$$V_{DS} = V_{DD} - I_D(R_D + R_S) \tag{7.23}$$

For an integrated circuit MOSFET, constant current sources are used for biasing. A basic MOSFET current source is shown in Figure 7.15.

For Figure 7.15, it can be shown that I_0 is related to I_{REF} through the expression

$$I_0 = \frac{\left(\frac{W}{L}\right)_2}{\left(\frac{W}{L}\right)_1} I_{REF} \tag{7.24}$$

and

$$I_{REF} = \frac{V_{DD} - V_{GS}}{R} \tag{7.25}$$

where

$(W/L)_2$ is ratio of width to length of transistor Q2

$(W/L)_1$ is ratio of width to length of transistor Q1

Thus, the current I_0 depends on the transistor sizing. The following example explores the changes in the drain current as the source resistor is changed.

Example 7.5   Effect of Source Resistance on MOSFET Operating Point

For the MOSFET biasing circuit shown in Figure 7.12, $V_{DD} = 10$ V, $R_{G1} = R_{G2} =$ 9 MΩ, and $R_D = 8$ KΩ. Find the drain current when R_S is varied from 5 KΩ to 10 KΩ in steps of 1 KΩ. Assume that M1 is M2N4351.

Solution

PSPICE is used to obtain the drain current as source current is varied. The **.STEP** command is used to vary the source resistance.

PSPICE program:

```
MOSFET BIAS CIRCUIT
VDD 4   0   DC 10V: SOURCE VOLTAGE
RG1 4   1   9.0E6;
RG2 1   0   9.0E6;
RD  4   3   8.0E3
M1  3   1   2   2   M2N4351; NMOS MODEL
RS  2   0   RMOD3 1
.MODEL RMOD3 RES(R=1)
.STEP RES RMOD3 (R) 5000 10000 1000: VARY RS FROM 5K TO 10K
.MODEL M2N4351 NMOS (KP=125U VTO=2.24 L=10U W=59U
LAMBDA=5M)
.DC VDD 10 10 1
.PRINT DC I(RD) V(3,2)
.END
```

TABLE 7.6

Drain Current vs. Source Resistance

Source Resistance R_S, Ω	Drain Current I_D, A
5000	3.576E–04
6000	3.094E–04
7000	2.731E–04
8000	2.447E–04
9000	2.218E–04
10K	2.030E–04

Table 7.6 shows the drain current for various values of source resistance. The data can also be found in file ex7_5ps.dat.

MATLAB is used to plot I_D vs. R_S.

MATLAB script:

```
% Load data
load 'ex7_5ps.dat' -ascii;
rs = ex7_5ps(:,1);
id = ex7_5ps(:,2);
plot(rs,id)
xlabel('Source Resistance')
ylabel('Drain Current')
title('Source Resistance vs. Drain Current')
```

Figure 7.16 shows the drain current as a function of source resistance. Figure 7.16 shows that as the source resistance increases, the drain current decreases. The following example explores the worst-case bias point with respect to device tolerance.

Example 7.6   Worst-Case Drain Current of a MOSFET Biasing Circuit

For the MOSFET biasing circuit shown in Figure 7.13, $V_{DD} = 5$ V, $V_{SS} = -5$ V, $R_G = 10$ MΩ, $R_S = R_D = 4$ KΩ. Determine the worst-case bias current when the tolerances of the resistors are 1%, 2%, 5%, 10%, and 15%, respectively. Assume that M1 is M2N4351.

Solution

The PSPICE **.WCASE** command is used to perform the worst-case analysis.

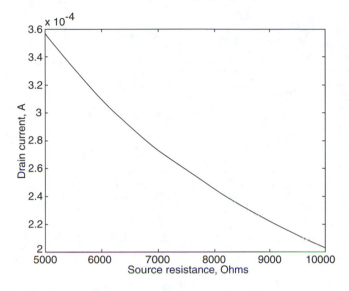

FIGURE 7.16
Drain current vs. source resistance.

PSPICE program:

```
* MOSFET BIASING CIRCUIT
.OPTIONS RELTOL = 0.01: 1% COMPONENT TOLERANCE,
* CHANGED FOR DIFFERENT TOLERANCE VALUES
VSS 5  0   DC  -5V
VDD 4  0   DC  5V
RG  1  0   RMOD 10.0E6
RS  2  5   RMOD 4.0E3
RD  4  3   RMOD 4.0E3
.MODEL RMOD RES(R=1 DEV=1%): 1% RESISTOR TOLERANCE.
* CHANGE FOR DIFFERENT TOLERANCE VALUES
M1 3 1 2 2 M2N4351
.MODEL M2N4351 NMOS (KP=125U VTO=2.24 L=10U W=59U
LAMBDA=5M)
.DC VDD 5 5 1
.WCASE DC I(RD) MAX OUTPUT ALL: WORST CASE ANALYSIS
.END
```

TABLE 7.7

Device Tolerance vs. Worst-Case
Drain Current

Device Tolerance (%)	Worst-Case (All Devices) Drain Current (A)
0	425.71E–06
1	428.98E–06
2	432.31E–06
5	442.62E–06
10	460.73E–06
15	480.81E–06

To obtain results for the 2% component tolerance, the two relevant statements in the above PSPICE program are changed to

.OPTION RELTOL = 0.02; 2% component tolerance
.MODEL RMOD RES(R = 1 DEV = 2%)

The above statements are appropriately changed for 5%, 10%, and 15% component tolerance.

Table 7.7 shows the worst-case (all devices) drain current vs. component tolerance in percent. The data can be found in file ex7_6ps.dat.

MATLAB is used to plot the data in Table 7.7.

MATLAB script:

```
% Load data
load 'ex7_6ps.dat' -ascii;
tol = ex7_6ps(:,1);
id_wc = ex7_6ps(:,2);
% plot data
plot(tol, id_wc, tol,id_wc,'ob')
xlabel('Device Tolerance, %')
ylabel('Worst-case Drain Current, A')
title('Worst-case Drain Current as a Function of Device
Tolerance')
```

The plot is shown in Figure 7.17.

Figure 7.17 shows that the worst-case drain current increases as the device tolerance increases.

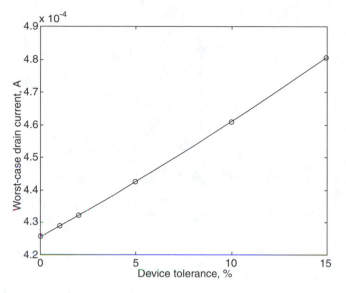

FIGURE 7.17
Device tolerance vs. worst-case drain current.

7.5 Frequency Response of Transistor Amplifiers

Amplifiers are normally used for voltage amplification, current amplification, impedance matching, or isolation between stages. Transistor amplifiers can be built using bipolar junction transistors (BJTs) and/or field effect transistors (FETs). Amplifiers built using BJTs can be common-emitter, common-collector (emitter follower), or common-base. Common-emitter amplifiers have a relatively high voltage gain. A common-collector amplifier has relatively high input resistance, low output resistance, with voltage gain that is almost equal to unity. Common-drain amplifiers have relatively low input resistance. FET amplifiers can be common-source, common-drain, or common-drain common-gate amplifier.

A common-emitter amplifier is shown in Figure 7.18. The amplifier is capable of generating relatively high current and voltage gains. The input resistance is medium and is essentially independent of the load resistance R_L.

The coupling capacitor C_{C1} couples the voltage source V_S to the bias network. Coupling capacitor C_{C2} connects the collector resistance R_C to the load R_L. The bypass capacitance C_E is used to increase the midband gain because it effectively short-circuits the emitter resistance R_E at midband frequencies. The resistance R_E is needed for bias stability. The external capacitors C_{C1}, C_{C2}, and C_E will influence the low-frequency response of the common-emitter

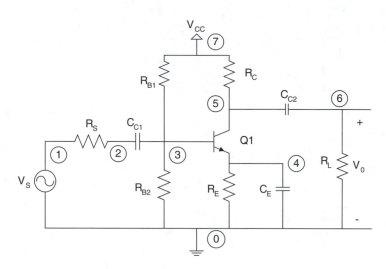

FIGURE 7.18
Common-emitter amplifier.

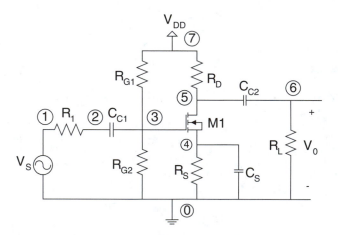

FIGURE 7.19
Common-source amplifier.

amplifier. The internal capacitances of the transistor will control the high-frequency cut-off.

The common-source amplifier shown in Figure 7.19 has characteristics similar to those of the common-emitter amplifier. However, the common-source amplifier has higher input resistance than that of the common-emitter amplifier.

The external capacitors C_{C1}, C_{C2}, and C_S will influence the low-frequency response. The internal capacitances of the FET will affect the high-frequency response. In the following example, the gain and bandwidth, as a function of power supply voltage, of the common-source amplifiers are explored.

Example 7.7 Common-Source Amplifier Characteristics

In the common-source amplifier shown in Figure 7.19, $C_{C1} = C_{C2} = 0.05$ µF, $C_S = 1000$ µF, $R_D = 6$ KΩ, $R_L = 10$ KΩ, $R_S = 2$ KΩ, $R_1 = 50$ Ω, $R_{G1} = 10$ MΩ, and $R_{G2} = 10$ MΩ. The MOSFET is IRF15O. Determine the midband gain, low cut-off frequency, and bandwidth as the power supply V_{DD} varies from 6 V to 10 V.

Solution

PSPICE is used to obtain the frequency response as the power supply is varied. The PSPICE command **.STEP** is used to vary the power supply voltage.

PSPICE program:

```
*  COMMON-SOURCE AMPLIFIER
VDD 7   0    DC 8V
.STEP   VDD 6   10  1
R1   1  2    50
CC1 2   3    0.05UF
RG2 3   0    10MEG
RG1 7   3    10MEG
RS   4  0    2K
CS   4  0    1000UF
RD   7  5    6K
VS   1  0    AC  1MV
CC2 5   6    0.05UF
RL   6  0    10K
M1   5  3    4   4   IRF150
.LIB NOM.LIB;
*  IRF 150 MODEL IN PSPICE LIBRARY FILE NOM.LIB
*  AC ANALYSIS
.AC DEC 20 10 10MEGHZ
.PRINT AC VM(6)
.PROBE V(6)
.END
```

PSPICE results for supply voltages of 6, 7, 8, 9, and 10 V are stored in files ex7_7aps.dat, ex7_7bps.dat, ex7_7cps.dat, ex7_7dps.dat, and ex7_7eps.dat, respectively. MATLAB is used to analyze the PSPICE results and to plot the frequency response.

MATLAB script:

```
% Load data
load 'ex7_7aps.dat' -ascii;
load 'ex7_7bps.dat' -ascii;
load 'ex7_7cps.dat' -ascii;
load 'ex7_7dps.dat' -ascii;
load 'ex7_7eps.dat' -ascii;
%
fre = ex7_7aps(:,1);
vo_6V = 1000*ex7_7aps(:,2);
vo_7V = 1000*ex7_7bps(:,2);
vo_8V = 1000*ex7_7cps(:,2);
vo_9V = 1000*ex7_7dps(:,2);
vo_10V = 1000*ex7_7eps(:,2);

% Determine center frequency
[vc1, k1] = max(vo_6V)
[vc2, k2] = max(vo_7V)
[vc3, k3] = max(vo_8V)
[vc4, k4] = max(vo_9V)
[vc5, k5] = max(vo_10V)
%
fc(1) = fre(k1): % center frequency for VDD = 6V
fc(2) = fre(k2): % center frequency for VDD = 7V
fc(3) = fre(k3): % center frequency for VDD = 8V
fc(4) = fre(k4): % center frequency for VDD = 9V
fc(5) = fre(k5): % center frequency for VDD = 10V
% Calculate the cut-off frequencies
vgc1 = 0.707*vc1: % Gain at cut-off for VDD = 6V
vgc2 = 0.707*vc2: % Gain at cut-off for VDD = 7V
vgc3 = 0.707*vc3: % Gain at cut-off for VDD = 8V
vgc4 = 0.707*vc4: % Gain at cut-off for VDD = 9V
vgc5 = 0.707*vc5: % Gain at cut-off for VDD = 10V
%
tol = 1.0e-5:   % tolerance for obtaining cut-off
```

(continued)

```
%
l1 = k1;
while(vo_6V(l1) - vgc1) > tol
     l1 = l1 +1;
end
fhi(1) = fre(l1):  % high cut-off frequency for VDD = 6V
l1 = k1
while(vo_6V(l1) - vgc1) > tol
     l1 = l1 - 1;
end
flow(1) = fre(l1):  % Low cut-off frequency for VDD = 6V
%
l2 = k2;
while(vo_7V(l2) - vgc2) > tol
     l2 = l2 + 1;
end
fhi(2) = fre(l2):  % high cut-off frequency for VDD = 7V
l2 = k2;
while(vo_7V(l2) - vgc2) > tol
     l2 = l2 - 1;
end
flow(2) = fre(l2):  % Low cut-off frequency for VDD = 7V
%
l3 = k3;
while(vo_8V(l3) - vgc3)>tol;
     l3 = l3 + 1;
end
fhi(3) = fre(l3):  % High cut-off frequency for VDD = 8V
l3 = k3
while(vo_8V(l3) - vgc3) > tol;
     l3 = l3 - 1;
end
flow(3) = fre(l3):  %low cut-off frequency for VDD = 8V
%
l4 = k4;
```

(continued)

```
while(vo_9V(14) - vgc4)>tol;
    14 = 14 + 1;
end
fhi(4) = fre(14):  % High cut-off frequency for VDD = 9V
14 = k4
while(vo_9V(14) - vgc4) > tol;
    14 = 14 - 1;
end
flow(4) = fre(14):  %low cut-off frequency for VDD = 9V
%
15 = k5;
while(vo_10V(15) - vgc5)>tol;
    15 = 15 + 1;
end
fhi(5) = fre(15):  % High cut-off frequency for VDD = 10V
15 = k5
while(vo_10V(15) - vgc5) > tol;
    15 = 15 - 1;
end
flow(5) = fre(15):  %low cut-off frequency for VDD = 10V
%
% Calculate the Quality Factor
for i = 1:5
bw(i) = fhi(i)-flow(i);
Qfactor(i) = fc(i)/bw(i);
end
%midband gain
gain_mb = [vc1 vc2 vc3 vc4 vc5];
% Print out results
% Gain Center frequency, high cut-off freq, low cut-off freq
and Q factor are
gain_mb
flow
bw
Qfactor
% plot frequency response
```

(continued)

```
plot(fre,vo_6V, fre,vo_7V, fre,
vo_8V,fre,vo_9V,fre,vo_10V)
xlabel('Frequency, Hz')
ylabel('Gain')
title('Frequency Response of a Common-source Amplifier')
```

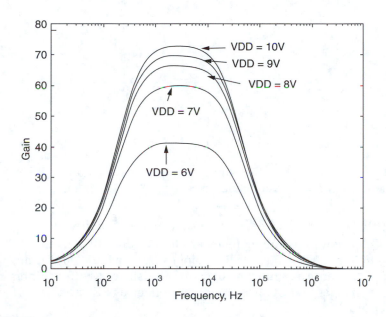

FIGURE 7.20
Frequency response of a common-source amplifier at different supply voltages.

The gain, low cut-off frequency, and bandwidth are shown in Figure 7.20. Table 7.8 shows the results obtained from MATLAB.

TABLE 7.8

Gain, Low Cut-off Frequency, and Bandwidth as a Function of Supply Voltage

Supply Voltage (V)	Midband Gain	Low Cut-off Frequency (Hz)	Bandwidth (Hz)
6	41.29	223.9	3.5256E+04
7	59.84	251.2	3.9559E+04
8	66.40	251.2	3.9559E+04
9	70.21	251.2	3.9559E+04
10	72.79	281.8	3.5198E+04

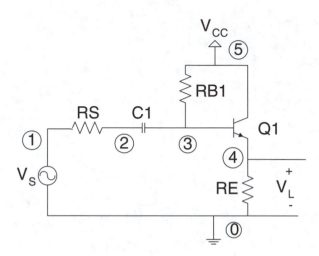

FIGURE 7.21
Emitter follower circuit.

Table 7.8 shows that as the supply voltage increases, the midband gain and the low cut-off frequency increase.

Example 7.8   Input Resistance of Emitter Follower

Figure 7.21 shows an emitter follower circuit. RS = 100 Ω, RB1 = 80 KΩ, VCC = 15 V, and C1 = 5 μF. If RE changes from 500 to 2000 Ω, determine the input resistance as a function of emitter resistance RE. In addition, determine the changes in the collector-emitter voltage as RE varies. Assume that Q1 is Q2N2222.

Solution

The **.STEP** statement is used to vary the emitter resistance RE in the PSPICE program.

PSPICE program:

```
*  INPUT RESISTANCE OF AN EMITTER FOLLOWER
VS   1   0   AC 10E-3
VCC  5   0   DC 15V
RS   1   2   100
C1   2   3   5UF
RB   5   3   80K
RE   4   0   RMOD4 1
.MODEL RMOD4 RES(R=1)
```

(continued)

```
.STEP RES RMOD4(R) 500 2000 150; VARY RE FROM 500 TO 2000
Q1  5  3  4   Q2N2222
.MODEL Q2N2222 NPN(BF=100 IS=3.295E-14 VA=200); TRANSISTOR
MODEL
.DC VCC 15 15 1
.AC LIN 1  1000  1000
.PRINT DC  I(RE) V(5,4)
.PRINT AC  V(1)  I(RS)
.END
```

The results obtained from PSPICE are shown in Table 7.9. The complete results are also available in file ex7_8ps.dat. MATLAB is used obtain the input impedance and also to plot the results.

TABLE 7.9

DC Emitter Current, DC Collector-Emitter Voltage, and Small-Signal Input Current vs. Emitter Resistance

Emitter Resistance (Ω)	DC Emitter Current (A)	DC Collector-to– Emitter Voltage (V)	Small-Signal Input Current (A)
500	1.136E–02	9.318	3.181E–07
650	1.013E–02	8.413	2.750E–07
800	9.147E–03	7.683	2.478E–07
950	8.338E–03	7.079	2.290E–07
1100	7.662E–03	6.572	2.153E–07
1250	7.088E–03	6.140	2.048E–07
1400	6.595E–03	5.767	1.965E–07
1550	6.166E–03	5.442	1.898E–07
1700	5.791E–03	5.156	1.842E–07
1850	5.458E–03	4.902	1.795E–07
2000	5.162E–03	4.675	1.755E–07

MATLAB script:

```
% Load data
load 'ex7_8ps.dat' -ascii;
re = ex7_8ps(:,1);
ie_dc = ex7_8ps(:,2);
vce_dc = ex7_8ps(:,3);
ib_ac = ex7_8ps(:,4);
vs_ac = 10.0e-03; % input signal is 10 mA
% Calculate input resistance
```

(continued)

```
m = length(re);
for i = 1:m
    rin(i) = vs_ac/ib_ac(i);
end
subplot(211), plot(re, rin, re, rin,'ob')
ylabel('Input Resistance, Ohms')
title('(a) Input Resistance')
subplot(212), plot(re, vce_dc, re, vce_dc, 'ob')
ylabel('Collector-emitter Voltage, V')
title('(b) DC Collector-Emitter Voltage')
xlabel('Emitter Resistance, Ohms')
```

The input resistance and collector-emitter voltage are shown in Figure 7.22.

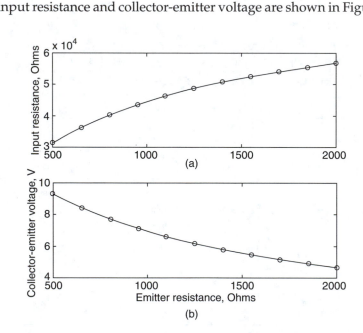

FIGURE 7.22
(a) Input resistance vs. emitter resistance and (b) collector-emitter voltage vs. emitter resistance.

7.6   Feedback Amplifiers

The general structure of a feedback amplifier is shown in Figure 7.23.

A is the open-loop gain of the amplifier without feedback. Input X_i and output X_0 are related by

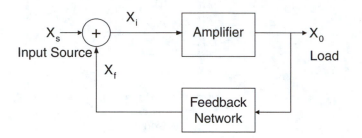

FIGURE 7.23
General structure of feedback amplifier.

$$X_0 = AX_i \qquad (7.26)$$

The output quantity X_0 is fed back to the input through the feedback network, which provides a sample signal X_f. The latter is related to the output X_0 by the expression

$$X_f = \beta X_0 \qquad (7.27)$$

The feedback signal X_f is subtracted from the source (for negative feedback amplifier) to produce the input signal

$$X_i = X_S - X_f \qquad (7.28)$$

For positive feedback, Equation (7.28) becomes

$$X_i = X_S + X_f \qquad (7.29)$$

Combining Equations (7.26) through (7.28), we get

$$A_f = \frac{x_O}{x_S} = \frac{A}{1+\beta A} \qquad (7.30)$$

where
    βA is the loop gain
    $(1 + \beta A)$ is the amount of feedback

It can be shown that amplifiers with negative feedback will result in (1) gain insensitivity to component variations, (2) increased bandwidth, and (3) reduced nonlinear distortion. The feedback amplifier topologies are shown in Figure 7.24.

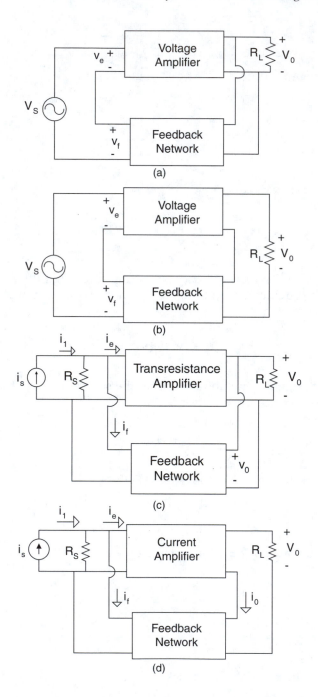

FIGURE 7.24
Feedback configurations: (a) series-shunt feedback, (b) series-series feedback, (c) shunt-resistor feedback, and (d) shunt-series feedback.

TABLE 7.10

Feedback Relationships

Amplifier Configuration	Gain	Input Resistance	Output Resistance
Without resistance	A	R_i	R_O
Series-shunt amplifier	$A_f = \dfrac{A}{1+\beta A}$	$R_{if} = R_i(1+\beta A)$	$R_{of} = \dfrac{R_O}{1+\beta A}$
Series-series amplifier	$A_f = \dfrac{A}{1+\beta A}$	$R_{if} = R_i(1+\beta A)$	$R_{of} = R_O(1+\beta A)$
Shunt-shunt amplifier	$A_f = \dfrac{A}{1+\beta A}$	$R_{if} = \dfrac{R_i}{1+\beta A}$	$R_{of} = \dfrac{R_O}{1+\beta A}$
Shunt-series amplifier	$A_f = \dfrac{A}{1+\beta A}$	$R_{if} = \dfrac{R_i}{1+\beta A}$	$R_{of} = R_O(1+\beta A)$

Depending on the type of feedback configuration, input and output resistance can be shown to increase or decrease by the feedback factor β. Table 7.10 shows the feedback relationships.

The following two examples will explore the characteristics of feedback amplifiers.

Example 7.9   Two-Stage Amplifier with Feedback Resistance

The circuit shown in Figure 7.25 is an amplifier with shunt-shunt feedback. RB1 = RB2 = 50 KΩ, RS = 100 Ω, RC1 = 5 KΩ, RE1 = 2.5 KΩ, RC2 = 10 KΩ,

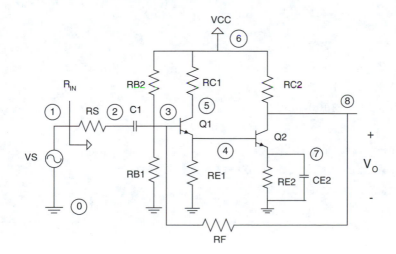

FIGURE 7.25
Amplifier with shunt-shunt feedback.

RE2 = 2 KΩ, C1 = 20 µF, CE2 = 100 µF, and VCC = 15 V. If VS = 1 mV, find V_0 as RF changes from 1 KΩ to 8 KΩ. Plot V_0 vs. RF. Assume that both transistors Q1 and Q2 are Q2N2222. The source voltage is a 2-KHz sine wave with 1-mV peak voltage.

Solution

PSPICE is used to obtain the output voltage with respect to RF. MATLAB is used to obtain the relationship between output voltage and RF.

PSPICE program:

```
AMPLIFIER WITH FEEDBACK
VS   1   0    AC 1MV 0
RS   1   2    100
C1   2   3    20E-6
RB1 3   0    50E3
RB2 6   3    50E3
RE1 4   0    2.5E3
RC1 6   5    5.0E3
Q1   5   3   4   Q2N2222
.MODEL Q2N2222 NPN(BF=100 IS=3.295E-14 VA=200):
TRANSISTORS MODEL
VCC 6   0    DC 15V
RE2 7   0    2E3
CE2 7   0    100E-6
RC2 6   8    10.0E3
Q2   8   4   7   Q2N2222
.AC LIN 1 2000 2000
RF 8 3 RMODF 1
.MODEL RMODF RES(R=1)
.STEP RES RMODF(R) 1.0E3 8.0E3 1.0E3
.PRINT AC VM(8,0)
.END
```

Table 7.11 shows the results obtained from PSPICE. The results are also available in file ex7_9ps.dat.

TABLE 7.11

Gain vs. Feedback Resistance

RF (Ω)	Gain
1000	7.932
2000	15.78
3000	23.38
4000	30.75
5000	37.91
6000	44.86
7000	51.62
8000	57.01

MATLAB is used to plot the results.

MATLAB script:

```
%load data
[rf, gain] = textread('ex7_9ps.dat', '%d %f');
plot(rf, gain, rf, gain,'ob')
title('Gain vs. Feedback Resistance')
xlabel('Feedback Resistance, Ohms')
ylabel('Gain')
```

The feedback resistance vs. gain is shown in Figure 7.26.

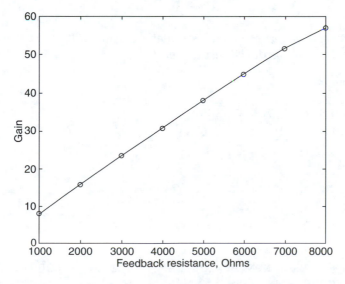

FIGURE 7.26

Feedback resistance vs. gain of a two-stage amplifier.

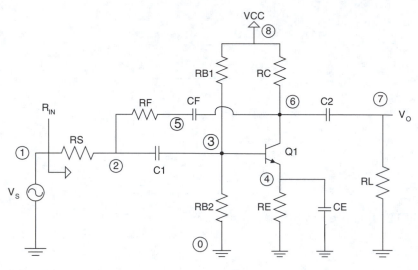

FIGURE 7.27
Common-emitter amplifier with shunt-shunt feedback.

Example 7.10 Common-Emitter Amplifier with Feedback Resistances

In Figure 7.27, RS = 150 Ω, RB2 = 20 KΩ, RB1 = 90 KΩ, RE = 2 KΩ, RC = 5 KΩ, RL = 10 KΩ, C1 = 2 μF, CE = 50 μF, C2 = 2 μF, CF = 5 μF, and VCC = 15 V. Find input resistance, R_{IN} and the voltage gain as the feedback resistance changes from 1 to 10 KΩ. The source voltage is a 1-KHz sine wave with 1-mV peak voltage. Assume that Q1 is Q2N2222.

Solution

PSPICE is used to perform circuit simulation. The PSPICE command **.STEP** is used to vary the feedback resistance.

PSPICE program:

```
COMMON EMITTER AMPLIFIER
VS   1   0   DC 0   AC 1E-30
RS   1   2   150
C1   2   3   2E-6
RB2  3   0   20E3
RB1  8   3   90E3
CF   5   6   5E-6
Q1   6   3   4 Q2N2222
.MODEL Q2N2222 NPN(BF=100 IS=3.295E-14 VA=200):
TRANSISTORS MODEL
```

(continued)

```
RE    4   0    2.0E3
CE    4   0    50E-6
RC    8   6    5.0E3
C2    6   7    2.0E-6
RL        7        0        10.0E3
VCC  8   0    DC   15V
RF        2   5    RMODF  1
.MODEL  RMODF  RES(R=1)
.STEP  RES  RMODF  (R)  1.0E3  10E3  1.0E3;
.AC  LIN  1  1000  1000
.PRINT  AC  I(RS)  V(7)
.END
```

Table 7.12 shows the PSPICE results. These are also available in file ex7_10ps.dat.

TABLE 7.12

Input Current and Output Voltage
as a Function of Feedback Resistance

Feedback Resistance (Ω)	Input Current (ac) (A)	Output Voltage (ac) (V)
1000	5.430E–06	5.169E–03
2000	5.208E–06	1.001E–02
3000	5.004E–06	1.445E–02
4000	4.817E–06	1.852E–02
5000	4.644E–06	2.228E–02
6000	4.484E–06	2.576E–02
7000	4.336E–06	2.898E–02
8000	4.198E–06	3.197E–02
9000	4.070E–06	3.477E–02
10000	3.950E–06	3.738E–02

MATLAB is used to calculate the input resistance, voltage gain, and also to plot the results.

MATLAB script:

```
% Load data
load 'ex7_10ps.dat' -ascii;
rf = ex7_10ps(:,1);
ib_ac = ex7_10ps(:,2);
```

(continued)

```
vo_ac = ex7_10ps(:,3);
vin_ac = 1.0e-3; % vs is 1 mA
% Calculate the input resistance and gain
n = length(rf); % data points in rf
for i = 1:n
    rin(i) = vin_ac/ib_ac(i);
    gain(i) = vo_ac(i)/vin_ac;
end
%
% Plot input resistance and gain
%
subplot(211)
plot(rf, rin, rf, rin,'ob')
title('(a) Input Resistance vs. Feedback Resistance')
ylabel('Input Resistance, Ohms')
subplot(212)
plot(rf, gain,rf,gain,'ob')
title('(b) Amplifier Gain vs. Feedback Resistance')
ylabel('Gain')
xlabel('Feedback Resistance, Ohms')
```

The input resistance and gain as a function of feedback resistance are shown in Figure 7.28.

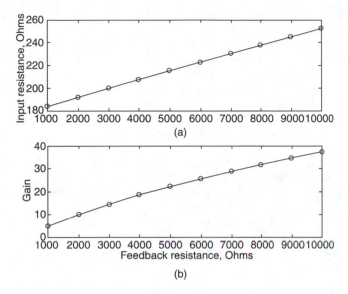

FIGURE 7.28
(a) Input resistance and (b) gain as a function of feedback resistance.

Bibliography

1. Al-Hashimi, Bashir, *The Art of Simulation Using PSPICE, Analog, and Digital*, CRC Press, Boca Raton, FL, 1994.
2. Attia, J.O., *Electronics and Circuit Analysis Using MATLAB*, CRC Press, Boca Raton, FL, 1999.
3. Biran, A. and Breiner, M., *MATLAB for Engineers*, Addison-Wesley, Reading, MA, 1995.
4. Boyd, Robert R., *Tolerance Analysis of Electronic Circuits Using MATLAB*, CRC Press, Boca Raton, FL, 1999.
5. Chapman, S. J., *MATLAB Programming for Engineers*, Brook, Cole Thompson Learning, Pacific Grove, CA, 2000.
6. Connelly, J. Alvin and Choi, Pyung, *Macromodeling with SPICE*, Prentice-Hall, Englewood Cliffs, NJ, 1992.
7. Distler, R.J., Monte Carlo Analysis of System Tolerance, *IEEE Trans. Education*, 20, 98–101, May 1997.
8. Ellis, George, Use SPICE to Analyze Component Variations in Circuit Design, *EDN*, pp. 109–114, April 1993,.
9. Etter, D.M., *Engineering Problem Solving with MATLAB*, 2nd edition, Prentice-Hall, Upper Saddle River, NJ, 1997.
10. Etter, D.M., Kuncicky, D.C., and Hull, D., *Introduction to MATLAB 6*, Prentice-Hall, Upper Saddle River, NJ, 2002.
11. Fenical, L. H., *PSPICE: A Tutorial*, Prentice-Hall, Englewood Cliffs, NJ, 1992.

12. Gottling, J.G., *Matrix Analysis of Circuits Using MATLAB*, Prentice-Hall, Englewood Cliffs, NJ, 1995.
13. Hamann, J.C., Pierre, J.W., Legowski, S.F., and Long, F.M., Using Monte Carlo Simulations to Introduce Tolerance Design to Undergraduates, *IEEE Trans. Education*, 42(1), 1–14, Feb. 1999.
14. Howe, Roger T. and Sodini, Charles G., *Microelectronics, An Integrated Approach*, Prentice-Hall, Upper Saddle River, NJ, 1997.
15. Kavanaugh, Michael F., Including the Effects of Component Tolerances in the Teaching of Courses in Introductory Circuit Design, *IEEE Trans. Education*, 38(4), 361–364, Nov. 1995.
16. Kielkowski, Ron M., *Inside SPICE, Overcoming the Obstacles of Circuit Simulation*, McGraw-Hill, New York, 1994.
17. Lamey, Robert, *The Illustrated Guide to PSPICE*, Delmar Publishers, Albany, NY, 1995.
18. Nilsson, James W. and Riedel, Susan A., *Introduction to PSPICE*, Addison-Wesley, Reading, MA, 1993.
19. OrCAD PSPICE A/D, Users' Guide, November 1998.
20. Prigozy, Stephen, Novel Applications of PSPICE in Engineering, *IEEE Trans. Education*, 32(1), 35–38, Feb. 1989.
21. Rashid, Mohammad H., *Microelectronic Circuits, Analysis and Design*, PWS Publishing Company, Boston, MA, 1999.
22. Rashid, Mohammad H., *SPICE for Circuits and Electronics Using PSPICE*, Prentice-Hall, Englewood Cliffs, NJ, 1990.
23. Roberts, Gordon W. and Sedra, Adel S., *SPICE for Microelectronic Circuits*, Saunders College Publishing, Fort Worth, TX, 1992.
24. Sedra, A.S. and Smith, K.C., *Microelectronic Circuits*, 4th edition, Oxford University Press, New York, 1998.
25. Sigmor, K., *MATLAB Primer*, 4th edition, CRC Press, Boca Raton, FL, 1998.
26. Spence, Robert and Soin, Randeep S., *Tolerance Design of Electronic Circuits*, Imperial College Press, River Edge, NY, 1997.
27. Thorpe, Thomas W., *Computerized Circuit Analysis with SPICE*, John Wiley & Sons, New York, 1991.
28. Tuinenga, Paul W., *SPICE, A Guide to Circuit Simulations and Analysis Using PSPICE*, Prentice-Hall, Englewood Cliffs, NJ, 1995.
29. Using MATLAB, The Language of Technical Computing, Computation, Visualization, and Programming, Version 6, MathWorks, Inc., 2000.
30. Vladimirescu, Andrei, *The SPICE Book*, John Wiley & Sons, New York, 1994.

Problems

7.1 A circuit for determining the input characteristics of a BJT is shown in Figure P7.1. R1 = 1 Ω, R2 = 1 Ω, VCC = 10 V, and Q1 is Q2N2222. (a) Obtain the input characteristics (i.e., IB vs. V_{BE}) by varying IB from 1 μA to 9 μA in steps of 0.1 μA. (b) Obtain the input resistance R_{BE} as a function of IB.

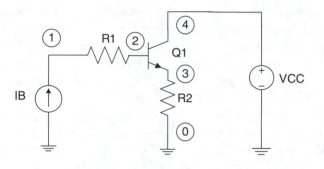

FIGURE P7.1
BJT circuit.

7.2 In Example 7.1, plot the output resistance, r_{CE}, as a function of V_{CE} for $I_B = 4\ \mu A$ and $6\ \mu A$. Is the output resistance dependent on I_B?

7.3 Figure 7.4 shows a configuration for obtaining MOSFET characteristics. (a) Obtain the output characteristics of MOSFET M2N4531 (i.e., I_{DS} vs. V_{DS}) for the following values of V_G: 3, 4, and 5 V. (b) For $V_{GS} = 4$ V, obtain the resistance, $r_{DS} = \Delta V_{DS}/\Delta I_{DS}$ for various values of V_{DS}. Plot r_{DS} vs. V_{DS}.

7.4 The data shown in Table P7.4 was obtained from a MOSFET. (a) Determine the threshold voltage V_T. (b) Determine the transconductance at $V_{GS} = 3$ V.

TABLE P7.4

I_{DS} vs. V_{GS} of a MOSFET

V_{GS} (V)	I_{DS} (mA)
1.0	3.375E–05
2.0	3.375E–05
2.8	3.375E–05
3.0	4.397E–02
3.2	2.093E–01
3.6	7.286E–01
4.0	7.385E–01
4.4	7.418E–01
4.8	7.436E–01

7.5 For the Widlar current source shown in Figure 7.9, $R_C = 20\ K\Omega$, $V_{CC} = 5$ V, and $R_E = 12\ K\Omega$. Determine the current I_o as a function of temperature (0°C to 120°C). Assume that the model for both Q1 and Q2 is

.MODEL npn_mod NPN(Bf = 150 Br = 2.0 VAF = 125V Is = 14fA Tf = 0.35ns Rb = 150 + Rc = 150 Re = 2 cje = 1.0pF Vje = 0.7V Mje = 0.33 Cjc = 0.3pF Vjc = 0.55V mjc = 0.5 + Cjs = 3.0pF Vjs = 0.52V Mjs = 0.5)

For resistances, assume that TC1 = 500 ppm/°C and T2 = 0.

7.6 In Example 7.3, determine the worst-case (for all devices) emitter current. Assume the tolerance of the resistors is 5%.

7.7 In Example 7.4, if Q1 is changed to Q2N2222 (model is available in PSPICE device library), and the temperature is varied from –25°C to 55°C, (a) plot emitter current vs. temperature, and (b) determine the best fit between the emitter current and the temperature.

7.8 For the MOSFET biasing circuit of Figure 7.14, determine the drain current as R_G takes the following values: 10^4, 10^5, 10^6, 10^7, 10^8, and $10^9 \, \Omega$. Assume that $V_{DD} = 15$ V, $R_D = 10$ KΩ, and transistor M1 is 1RF150 (model is available in PSPICE device library).

7.9 In Example 7.7, obtain the input resistance as a function of power supply voltage (7 V to 10 V) when the frequency of the source is 5000 Hz. Plot input resistance vs. supply voltage.

7.10 In the common-emitter amplifier shown in Figure 7.18, $C_{C1} = C_{C2} = 5 \, \mu F$, $C_E = 100 \, \mu F$, $R_{B1} = 50$ KΩ, $R_{B2} = 40$ KΩ, $R_S = 50 \, \Omega$, $R_L = R_C = 10$ KΩ, $R_E = 1$ KΩ. Transistor Q1 is Q2N3904. Determine (a) gain, (b) maximum input resistance, (c) low cut-off frequency, and (d) bandwidth as a function of supply voltage VCC (8 V to 12 V).

7.11 In Example 7.8, obtain the voltage gain as a function of the emitter resistance.

7.12 The circuit shown in Figure P7.12 is a Darlington amplifier. Q1 and Q2 are the Darlington-pair. The circuit has a very high input resistance. RB = 80 KΩ, RS = 100 Ω, C1 = 5 µF, VCC = 15 V, and transistors Q1 and Q2 are both Q2N2222. If RE varies from 500 Ω to 1500 Ω, determine the input resistance RIN and voltage gain as a function of emitter resistance. Assume that input voltage VS is a sinusoidal waveform with a frequency of 2 KHz and a peak value of 10 mV.

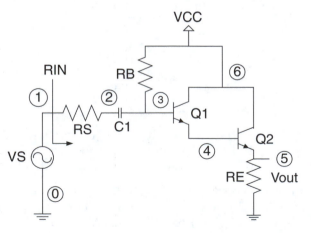

FIGURE P7.12
Darlington amplifier.

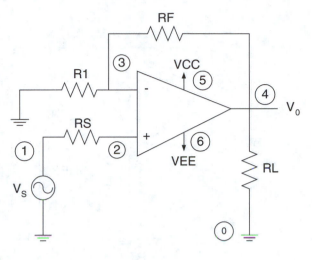

FIGURE P7.13
Op amp circuit with series-shunt feedback network.

7.13 Figure P7.13 is an op amp circuit with a series-shunt feedback network. RS = 1 KΩ, RL = 10 KΩ, and R1 = 5 KΩ. Find the gain, V_o/V_s if RF varies from 10 KΩ to 100 KΩ. Plot the voltage gain with respect to RF. Assume that the op amp is UA741 and the input voltage V_S is a sinusoidal waveform with a frequency of 5 KHz and a peak voltage of 1 mV.

7.14 A two-stage amplifier with shunt-series feedback is shown in Figure P7.14. RB1 = 60 KΩ, RB2 = 80 KΩ, RS = 100 Ω, RC1 = 8 KΩ, RE1 = 2.5 KΩ, RB3 = 50 KΩ, RB4 = 60 KΩ, RC2 = 5 KΩ, RE2 = 1 KΩ, C1 = 20 µF, CE = 100 µF, C2 = 20 µF, and VCC = 15 V. If input voltage VS is a sinusoidal voltage with a peak value of 1 mV and a frequency of 1 KHz, determine the input resistance R_{IN} and output voltage as RF varies from 1 KΩ to 6 KΩ. Assume that both transistors Q1 and Q2 are Q2N3904.

7.15 For Problem 7.14, determine the low cut-off frequency, high cut-off frequency, and the bandwidth as a function of the feedback resistance RF.

7.16 For the two-stage amplifier with shunt-series feedback shown in Figure P7.14, RB1 = 60 KΩ, RB2 = 80 KΩ, RS = 100 Ω, RC1 = 8 KΩ, RE1 = 2.5 KΩ, RB3 = 50 KΩ, RB4 = 60 KΩ, RC2 = 5 KΩ, RE2 = 1 KΩ, RF = 2 KΩ, C1 = 20 µF, CE = 100 µF, and C2 = 20 µF. If the input voltage VS is a sinusoidal voltage with a peak value of 1 mV, determine the voltage gain and bandwidth as a function of the supply voltage VCC (11 V to 15 V). Assume that both transistors Q1 and Q2 are Q2N3904.

7.17 For Example 7.9, if VS = 1 mV and frequency of the source is 5 KHz, find the input resistance as RF varies from 1 KΩ to 8 KΩ.

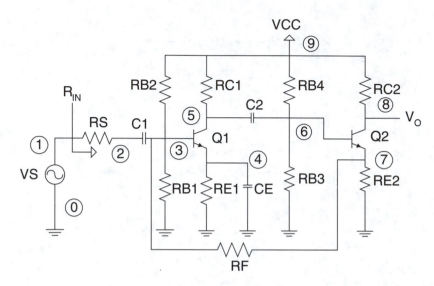

FIGURE P7.14
Two-stage amplifier with shunt-series feedback.

Index